Bacterial Sensors

Synthetic Design and Application Principles

Synthesis Lectures on Synthetic Biology

Editor
Martyn Amos, *Manchester Metropolitan University*

This series will publish 50- to 125-page publications on topics at the interface(s) of molecular and microbiology, engineering, computer science, physics, chemistry and mathematics. Potential topics include, but are not limited to: Bio-fabrication, Bio-interfacing, Computational bio-mimetics, Computational genes and molecular automata, DNA nanotechnology, Ethical, legal and social issues, Gene networks, Metabolic engineering, Modeling and simulation approaches, (Multi)-cellular computing/engineering, Nanobiotechnology, Protein engineering, Security, Self-assembly, and Standardization.

Bacterial Sensors: Synthetic Design and Application Principles
Jan Roelof van der Meer
2010

Genome Refactoring
Natalie Kuldell and Neal Lerner
2009

© Springer Nature Switzerland AG 2022

Reprint of original edition © Morgan & Claypool 2011

Bacterial Sensors: Synthetic Design and Application Principles

Jan Roelof van der Meer

ISBN: 978-3-031-01442-0   paperback
ISBN: 978-3-031-02570-9   ebook

DOI 10.1007/978-3-031-02570-9

A Publication in the Springer series
SYNTHESIS LECTURES ON SYNTHETIC BIOLOGY

Lecture #2
Series Editor: Martyn Amos, *Manchester Metropolitan University*
Series ISSN
Synthesis Lectures on Synthetic Biology
Print 2151-0008   Electronic 2151-0016

Bacterial Sensors

Synthetic Design and Application Principles

Jan Roelof van der Meer
University of Lausanne, Switzerland

SYNTHESIS LECTURES ON SYNTHETIC BIOLOGY #2

ABSTRACT

Bacterial reporters are live, genetically engineered cells with promising application in bioanalytics. They contain genetic circuitry to produce a cellular sensing element, which detects the target compound and relays the detection to specific synthesis of so-called reporter proteins (the presence or activity of which is easy to quantify). Bioassays with bacterial reporters are a useful complement to chemical analytics because they measure biological responses rather than total chemical concentrations. Simple bacterial reporter assays may also replace more costly chemical methods as a first line sample analysis technique. Recent promising developments integrate bacterial reporter cells with microsystems to produce bacterial biosensors.

This lecture presents an in-depth treatment of the synthetic biological design principles of bacterial reporters, the engineering of which started as simple recombinant DNA puzzles, but has now become a more rational approach of choosing and combining sensing, controlling and reporting DNA 'parts'. Several examples of existing bacterial reporter designs and their genetic circuitry will be illustrated. Besides the design principles, the lecture also focuses on the application principles of bacterial reporter assays. A variety of assay formats will be illustrated, and principles of quantification will be dealt with. In addition to this discussion, substantial reference material is supplied in various Annexes.

KEYWORDS

biosensor, bioreporter, synthetic biology, promoter engineering

Contents

Preface

Bacterial reporters are live, genetically engineered cells with promising application in bioanalytics. They contain genetic circuitry to produce a cellular sensing element, which detects the target compound and relays the detection to specific synthesis of so-called reporter proteins (the presence or activity of which is easy to quantify). Bioassays with bacterial reporters are a useful complement to chemical analytics because they measure biological responses rather than total chemical concentrations. Simple bacterial reporter assays may also replace more costly chemical methods as a first line sample analysis technique. Recent promising developments integrate bacterial reporter cells with microsystems to produce bacterial biosensors.

This lecture presents an in-depth treatment of the synthetic biological design principles of bacterial reporters, the engineering of which started as simple recombinant DNA puzzles, but has now become a more rational approach of choosing and combining sensing, controlling and reporting DNA 'parts'. Several examples of existing bacterial reporter designs and their genetic circuitry will be illustrated. Besides the design principles, the lecture also focuses on the application principles of bacterial reporter assays. A variety of assay formats will be illustrated, and principles of quantification will be dealt with. In addition to this discussion, substantial reference material is supplied in various Annexes.

Jan Roelof van der Meer
November 2010

Acknowledgments

This work would not have been possible without the dedicated research and support of a number of PhD students and postdocs in my laboratory, most notably Marco Jaspers (who led the basis for much of our bioreporter work and was responsible for compiling most of the data in the Annexes 1-3), Judith Stocker, Barbara Baumann and Patrick Sticher, Robin Tecon (who was instrumental in laying the basis for the bioreporter flux studies), David Tropel, Siham Beggah and Christel Vogne (exploring the possibilities for sensory protein mutagenesis), Rekha Kumari, and more recently Davide Merulla and Nina Buffi (for their enthusiasm in developing microengineering tools for bioreporters), and Artur Reimer (to explore periplasmic binding proteins as analyte sensors). I am also much indebted to my past and present collaborators Mona Wells and Hauke Harms to mature the arsenic bioreporter assay, Victor de Lorenzo (with his vast knowledge, experience and ideas in developing cloning tools), and Shimshon Belkin, world-leading expert in bacterial bioreporters for toxicity. Funding support from the Swiss National Science Foundation program Sinergia, and from the European Framework programmes (projects BIOCARTE, FACEiT, BACSIN and BIOMONAR) is widely acknowledged.

Finally, I apologize to any authors and groups whose work is not or underrepresented in this lecture, which is – despite my attempts to make it as broad as possible, still a limited overview from my own perception. I will be glad to improve and correct the text upon suggestions by readers.

Jan Roelof van der Meer
November 2010

CHAPTER 1

Short History of the use of Bacteria for Biosensing and Bioreporting

1.1 EARLY WARNING SYSTEMS

Mankind has long realized that potentially harmful toxic compounds can be present in food, water or air around them, and has employed living organisms other than himself to analyze this risk. The saying 'canary in a coal mine' is a reflection on the use of canaries to probe the presence of carbon monoxide and methane in mine shafts, which was common practice in Britain from 1911, and even as recently as 1986[1]. Mine workers used to bring canaries in a hand-held cage down into the shafts and listen to their singing, which would interrupt or fade in case of the birds being exposed to noxious mine gases. This 'acoustic' signaling would then give the workers sufficient time to leave the dangerous zones. The basis for the feasibility of the concept of using canaries for early warning is that the birds are more sensitive to mine gases than humans, which is assumed to be caused by a unique different design of the avian respiratory system [Brown et al., 1997]. Importantly, the birds were not used as replacement for human physiological behavior, but for discrimination of the presence or absence of hazard. The use of animals has been and still is very common; most notably in toxicological and pharmaceutical (drug, cosmetics) tests, a practice fiercely disputed when it concerns mammals and other higher animals. This dispute not only occurs on ethical grounds, but also for reasons that some authors describe as the 'high fidelity fallacy', the idea that 'a (testing) model (must) resemble the target in every aspect'[Festing, F., 2009]. At the level of environmental safety, animals also continue to be regularly employed both as continuous early-warning systems - for instance, in drinking water intake (fish, daphnia) - and as biomarkers for in-situ or offline exposure tests (nematodes, worms, crabs or clams) [Lewis et al., 2008]. As a consequence of this practice, numerous methods for the replacement of unnecessary toxicological and ecotoxicological testing have been proposed. Many such assays are based on single cell microorganisms; others are on mammalian, avian and fish cell lines [Eggen and Segner, 2003]. Such biological alternatives are clearly not high fidelity models of human physiological behavior but are models that may offer an appropriate 'discrimination between compounds which are toxic and non-toxic at a given dose level' [Festing, F., 2009]. The debate, however, over whether microorganisms or cell lines need to be

---

[1](http://news.bbc.co.uk/onthisday/hi/dates/stories/december/30/newsid_2547000/2547587.stm)

high fidelity models for human physiology or rather discriminatory tests, and as such can replace costly and morally unjust animal tests, is far from over. Fortunately, many authors and regulatory agencies are becoming more amenable to the idea of replacing animal models with microbial and cell line test systems[2]. Microbial-based reporter test systems would be a particularly attractive option because of their general amenability for genetic manipulation and systems design.

1.2   EARLY USE OF BACTERIAL 'BIOREPORTERS'

Bacteria have established their position as systems for testing dose-dependent toxicity or mutagenicity, although it is quite well realized that bacteria are not *per se* accurate models of human physiology. On the other hand, even the simplest response pathways in bacteria may give appropriate indications for the potential toxicity of a compound. The first of such tests was developed in the late 1960s and early seventies and became known as the Ames-test or 'mutatest'. The basis of the test is a mutagen-induced back-reversion of an auxotrophic mutation in the histidine biosynthesis operon of *Salmonella typhimurium* LT-2 [Ames et al., 1973b]. The assay consists of growing the test strain on agar media with or without the mutagenic compound, and scoring the numbers of colonies that develop. In the absence of mutagen, the test strain cannot grow to visible colony size because the medium contains only traces of histidine, which suffices for survival but does not permit millions of bacterial cells to propagate. However, if the test substance in the medium induces mutations in the *his* operon that lead to reversion of the auxotrophy, the bacteria *will* develop visible colonies. The number of colonies of the test strain on media with the mutagen is corrected for that on plates *without* the mutagen (those reflecting spontaneous back-reversion of the *his* marker), and then used as a proxy for the compound's mutagenicity. In order to detect a larger spectrum of mutagenic action (e.g., frameshift, base substitution), different *S. typhimurium his* auxotrophic mutants are used in separate test assays. The tester strains often also carry additional mutations to increase sensitivity to the potential mutagen. Typically, these are the *rfa* mutation, which diminishes lipopolysaccharide capsid synthesis and is thought to result in more easy compound entry. Secondly, *uvrAB* mutations are used, which prevent DNA repair systems [Ames et al., 1973b]. Apart from screening for back-reversion of the *his* auxotrophic marker, the tests were originally also designed to be used as 'repair-test', in which two strains, one with and one without DNA repair system, are compared for the mutagen-induced killing zone on agar plates with histidine [Ames et al., 1973b]. Further improvement of the test was proposed by including a pre-incubation of the test compound with rat-liver extract. The idea here was that enzymes in the extract would modify certain target compounds, which themselves had no mutagenic activity, to become mutagens in the test. Because rat-liver enzymes would mimic detoxification activity of the liver, the assay would permit to 'detect' more than the reaction of the bacterial cells themselves. The modified test was therefore claimed to detect potential 'carcinogenic' compounds in addition to purely mutagenic ones [Ames et al., 1973a]. This concept has recently again

---

[2]   (http://ec.europa.eu/enterprise/epaa/index_en.htm)   (http://imi.europa.eu/docs/ imi-gb-006v2-15022008-research-agenda_en.pdf)   (http://www.europarl.europa.eu/code/dossier/ 2002/2000_0077_cosmetic/cosmetic_table_en.pdf)   http://ec.europa.eu/environment/chemicals/reach/ reach_intro.htm. Consulted on August 21 2009.

been taken up by expressing human P450 enzymes directly in bacterial reporter cells [Biran et al., 2009].

Whereas the Ames-test has been extremely successful, it can not really be claimed to be a gene reporter assay. The 'honor' for being the first gene reporter assay is likely reserved for the SOS chromotest, which was developed in 1982 by Quillardet et al. [1982]. Like the Ames-test, the SOS chromotest also screens for potential compound mutagenicity but employs the SOS response of *E. coli* cells induced by DNA damage. SOS response is measured via expression of *lacZ* (for beta-galactosidase), which for this purpose was fused to the *sfiA* gene (now called *sulA*). The *sfiA* gene, the product of which is implicated in cell division control, was at that point identified as one of the genes strongly induced during SOS response [Huisman and D'Ari, 1981]. In order to optimize detection of beta-galactosidase induced by mutagens the *E. coli* tester strain with the *sfiA::lacZ* fusion was deleted for the natural *lac* operon. Like the *S. typhimurium* used for the Ames-test, the SOS test strain was also made deficient for UvrA-mediated excision repair and polysaccharide capsid synthesis (*raf*) [Quillardet et al., 1982]. Because the assay output relied on *de novo* protein synthesis, any substance influencing protein synthesis rather than DNA damage per se would lead to an underestimation in the test. For this reason, the tester strain also constitutively produces alkaline phosphatase activity, and it is the ratio between beta-galactosidase and alkaline phosphatase that is measured in the assay [Quillardet et al., 1982]. The final quantitative value of the assay is then derived as the slope of the linear portion of a dose-response curve for each target compound. Results from Ames- and SOS chromotests generally correlate quite well, with the Ames test being slightly more sensitive to mutagenic compounds but the SOS chromotest producing fewer false-positives, and both tests are still widely in use [Biran et al., 2009, Quillardet and Hofnung, 1993, Reifferscheid and Hell, 1996].

Along similar lines, a further bioreporter test was proposed, which uses the *umuDC* operon from *E. coli*, encoding the UmuDC error-prone DNA repair system [Oda et al., 1985]. The expression of the *umu* operon in *E. coli* is also induced by the SOS response, elicited by chemicals or radiation that damage DNA [Fry et al., 2005]. The constructed reporter system consisted of a *S. typhimurium* strain containing a plasmid (pSK1002) with a *umuC'-'lacZ* translational fusion. As before, mutagenic and carcinogenic chemicals that cause DNA damage would thus induce beta-galactosidase expression. This test, which became known as the *umu*-test, provided a very rapid and practical assay to screen chemicals for genotoxic effects, and, in contrast to the Ames test, did not require multiple tester strains simultaneously. Comparative studies have shown that the same compounds produce quite similar reactions in Ames, SOS chromo- and *umu*-test [Oda et al., 1985, Reifferscheid and Hell, 1996]. The test has been made more sensitive by deleting the natural *lac* operon in the strain and by producing the same mutations in UvrB repair and *rfa* as before in the Ames and SOS chromotest strains. The *umu*-test is so far the only bacterial bioreporter assay that has obtained international method standardization according to the DIN (Deutsche Industrie Norm, DIN 38415-3) and ISO (International Standard Organisation, ISO/CD 13829) guidelines.

Whereas the Ames-, *umu* and SOS chromotests have a strong emphasis on the detection of mutagenic compounds, other bacterial-based assays appeared, which coupled distress caused to the bacterial cell by chemicals or mixtures of chemicals to changes in general physiological behavior or enzymatic activities [Bitton and Koopman, 1992]. Such general physiological responses include, for example, oxygen respiration rate, cytoplasmic esterase activity, or chemotaxis rates [Bitton and Koopman, 1992]. One of the physiological tests became known as the Microtox^{TM}-test and was originally proposed by Bulich and Isenberg [1981], although earlier mention for the use of bioluminescence in toxicity screening was made [Ulitzur and Goldberg, 1977]. The Microtox^{TM}-test is an assay in which the marine bioluminescent bacterial species *Vibrio fischeri* (now officially named *Allovibrio fischeri*) is incubated under controlled medium and temperature conditions with test samples or chemicals while constantly measuring the bioluminescence activity. Because bioluminescence is a highly energy and cofactor demanding process for the cell, any compound perturbing this process will subside luciferase activity and as a consequence decrease light output from the cells in the assay. Numerous modifications of the original Microtox^{TM}-test have been proposed, such as employing more sensitive bioluminescent bacterial species like *Photobacterium leiognathi* [Ulitzur et al., 2002]. Test variations have become known under different trade names, like BioTox^{TM}, LUMIStox^{TM} or ToxAlert^{TM}. Although the Microtox^{TM}-test is not a bioreporter test in the sense of utilizing a synthetic reporter circuit design, it is and has been very widely and successfully applied, and many luciferase systems have found their use in synthetic bioreporters (see Chapter 2). Reporter test systems like Microtox^{TM} sometimes are called 'lights-off' systems because the effects of exposure of the cells to toxicants result in a decrease of a constitutively produced bioluminescence [Daunert et al., 2000]. In contrast, inducible reporters, like most of the ones described further in this text, would be called 'lights-on' systems because they only produce a reporter signal in the presence but not in the absence of the target compound.

Whereas the early toxicity and mutagenicity assays integrated an inducible reporter response into a wide toxicity network of the cell, the first 'compound-specific' bacterial reporter assay appeared in the early 1990s with the engineering of a naphthalene-sensing *Pseudomonas fluorescens* [King et al., 1990]. The naphthalene-sensing bacterium was constructed by bringing expression of bacterial luciferase under control of the same regulatory network that controls naphthalene metabolism. The result was an assay in which the bacteria produced bioluminescence as a function of naphthalene exposure, which could be easily and sensitively recorded at a distance without further manipulation of the cells [King et al., 1990].

While these examples demonstrated the potential usefulness of bioreporter bacteria for measuring toxic chemicals, they did not employ any specific rational design because of the lack of understanding of the molecular and regulatory basis of the used promoters and networks. Through advances over the past 10-15 years in molecular genetics, genomics and synthetic biology, a much better idea exist on the different levels of regulatory control in the cell, on the variety of sensory and regulatory proteins that bacteria posses, and on additional layers of regulatory complexity in the form of posttranscriptional and –translational mechanisms. Historically, and for reasons outlined above, there has been a strong focus on designing and constructing bacterial bioreporters, which detect chemical compounds of environmental relevance, toxic, or mutagenic compounds. More and more genetic circuits are being engineered for completely different reasons, such as study of noise in networks [Pedraza, P., 2005], production of oscillators [Elowitz and Leibler, 2000], cells that count [Friedland et al., 2009], or efficient coupling of metabolic networks [Keasling and Chou, 2008].

Various aspects of bioreporter design and use have been summarized in scholarly reviews, demonstrating the general and broad interest in the topic. Mentioning a few here may serve for further entry into the subject for the interested reader. The first broad review on bioreporters focuses on the various reporter genes and presents examples of reporter strains with applications as far as known at that time [Daunert et al., 2000]. Similarly, focused reviews have appeared over the course of the years [Belkin, B., 2003, Biran et al., 2009, D'Souza, D., 2001, Keane et al., 2002, Kohler et al., 2000, Ron, R., 2007, van der Meer and Belkin, 2010, Yagi, Y., 2007]. Other entry points for discussing bioreporters have included single cell applications [Tecon and van der Meer, 2006], specific applications of bioreporters to understand the ecological roles of microbes [Jansson, J., 2003, Leveau and Lindow, 2002], quantitative bioreporter measurements [van der Meer et al., 2004], specific focus on genotoxicity reporters only [Biran et al., 2009, Sorensen et al., 2006], on metal-responsive reporters only [Magrisso et al., 2008], on reporters for organic pollutants only [Keane et al., 2002, Yagi, Y., 2007], on the issues of compound bioavailability [Magrisso et al., 2008, Tecon and van der Meer, 2008], or on the questions why bioreporters are deemed so successful in laboratory research but not widely applied [Harms et al., 2006]. More practical aspects of bioreporter assaying have also passed the revue: formulation and maintenance of bioreporter cells [Bjerketorp et al., 2006, Marques et al., 2006], and, more recently, applications of bioreporter cells in micro-engineered systems [Nivens et al., 2004, van der Meer and Belkin, 2010]. The main purpose of this lecture is to focus on the design principles of bioreporter circuits (Chapter 2), and to present the basis of quantitative bioreporter assaying (Chapter 3).

REFERENCES

Ames, B. N., W. E. Durston, et al. (1973a). "Carcinogens are mutagens: a simple test system combining liver homogenates for activation and bacteria for detection." *Proc. Natl. Acad. Sci. USA* 70(8): 2281–2285. DOI: 10.1073/pnas.70.8.2281 2

Ames, B. N., F. D. Lee, et al. (1973b). "An improved bacterial test system for the detection and classification of mutagens and carcinogens." *Proc. Natl. Acad. Sci. USA* 70(3): 782–786. DOI: 10.1073/pnas.70.3.782 2

Belkin, S. (2003). "Microbial whole-cell sensing systems of environmental pollutants." *Curr. Opin. Microbiol.* 6(3): 206–212. DOI: 10.1016/S1369-5274(03)00059-6 5

Biran, A., R. Pedahzur, et al. (2009). Genetically engineered bacteria for genotoxicity assessment. *Biosensors for the Environmental Monitoring of Aquatic Systems.* D. Barcelo and P.-D. Hansen. Berlin/ Heidelberg, Springer: 161–186. DOI: 10.1007/978-3-540-36253-1_6 3, 5

Bitton, G. and B. Koopman (1992). "Bacterial and enzymatic bioassays for toxicity testing in the environment." *Rev. Environ. Contam. Toxicol.* 125: 1–22. 4

Bjerketorp, J., S. Hakansson, et al. (2006). "Advances in preservation methods: keeping biosensor microorganisms alive and active." *Curr. Opin. Biotechnol.* 17(1): 43–49. DOI: 10.1016/j.copbio.2005.12.005 5

Brown, R. E., J. D. Brain, et al. (1997). "The avian respiratory system: a unique model for studies of respiratory toxicosis and for monitoring air quality." *Environ. Health Perspect.* 105(2): 188–200. DOI: 10.2307/3433242 1

Bulich, A. A. and D. L. Isenberg (1981). "Use of the luminescent bacterial system for the rapid assessment of aquatic toxicity." *ISA Trans.* 20(1): 29–33. 4

D'Souza, S. F. (2001). "Microbial biosensors." *Biosens. Bioelectr.* 16: 337–353. DOI: 10.1016/S0956-5663(01)00125-7 5

Daunert, S., G. Barrett, et al. (2000). "Genetically engineered whole-cell sensing systems: coupling biological recognition with reporter genes." *Chem. Rev.* 100(7): 2705–2738. DOI: 10.1021/cr990115p 4, 5

Eggen, R. I. and H. Segner (2003). "The potential of mechanism-based bioanalytical tools in ecotoxicological exposure and effect assessment." *Anal. Bioanal. Chem.* 377: 386–396. DOI: 10.1007/s00216-003-2059-y 1

Elowitz, M. B. and S. Leibler (2000). "A synthetic oscillatory network of transcriptional regulators." *Nature* 403(6767): 335–338. DOI: 10.1038/35002125 5

Festing, M. F. (2009). "Fifty years after Russell and Burch, toxicologists continue to ignore genetic variation in their test animals." *Altern. Lab. Anim.* 37(1): 1–5. 1

Friedland, A. E., T. K. Lu, et al. (2009). "Synthetic gene networks that count." *Science* 324(5931): 1199–1202. DOI: 10.1126/science.1172005 5

Fry, R. C., T. J. Begley, et al. (2005). "Genome-wide responses to DNA-damaging agents." *Annu. Rev. Microbiol.* 59: 357–377. DOI: 10.1146/annurev.micro.59.031805.133658 3

Harms, H., M. C. Wells, et al. (2006). "Whole-cell living biosensors–are they ready for environmental application?" *Appl. Microbiol. Biotechnol.* 70(3): 273–280. DOI: 10.1007/s00253-006-0319-4 5

Huisman, O. and R. D'Ari (1981). "An inducible DNA replication-cell division coupling mechanism in *E. coli.*" *Nature* 290(5809): 797–799. DOI: 10.1038/290797a0 3

Jansson, J. K. (2003). "Marker and reporter genes: illuminating tools for environmental microbiologists." *Curr. Opin. Microbiol.* 6: 310–316. DOI: 10.1016/S1369-5274(03)00057-2 5

Keane, A., P. Phoenix, et al. (2002). "Exposing culprit organic pollutants: a review." *J. Microbiol. Meth.* 49(2): 103–119. DOI: 10.1016/S0167-7012(01)00382-7 5

Keasling, J. D. and H. Chou (2008). "Metabolic engineering delivers next-generation biofuels." *Nat. Biotechnol.* 26(3): 298–299. DOI: 10.1038/nbt0308-298 5

King, J. M. H., P. M. DiGrazia, et al. (1990). "Rapid, sensitive bioluminescent reporter technology for naphthalene exposure and biodegradation." *Science* 249(4970): 778–781. DOI: 10.1126/science.249.4970.778 4

Kohler, S., S. Belkin, et al. (2000). "Reporter gene bioassays in environmental analysis." *Fresenius J. Anal. Chem.* 366(6–7): 769–779. DOI: 10.1007/s002160051571 5

Leveau, J. H. J. and S. E. Lindow (2002). "Bioreporters in microbial ecology." *Curr. Opin. Microbiol.* 5: 259–265. DOI: 10.1016/S1369-5274(02)00321-1 5

Lewis, C., C. Pook, et al. (2008). "Reproductive toxicity of the water accommodated fraction (WAF) of crude oil in the polychaetes *Arenicola marina* (L.) and *Nereis virens* (Sars)." *Aquatic Toxicol.* 90(1): 73–81. DOI: 10.1016/j.aquatox.2008.08.001 1

Magrisso, S., Y. Erel, et al. (2008). "Microbial reporters of metal bioavailability." *Microb. Biotechnol.* doi:10.1111/j.1751–7915.2008.00022.x. 5

Marques, S., I. Aranda-Olmedo, et al. (2006). "Controlling bacterial physiology for optimal expression of gene reporter constructs." *Curr. Opin. Biotechnol.* 17(1): 50–56. DOI: 10.1016/j.copbio.2005.12.001 5

Nivens, D. E., T. E. McKnight, et al. (2004). "Bioluminescent bioreporter integrated circuits: potentially small, rugged and inexpensive whole-cell biosensors for remote environmental monitoring." *J. Appl. Microbiol.* 96: 33–46. DOI: 10.1046/j.1365-2672.2003.02114.x 5

Oda, Y., S. Nakamura, et al. (1985). "Evaluation of the new system (*umu*-test) for the detection of environmental mutagens and carcinogens." *Mutat. Res.* 147(5): 219–229. 3

Pedraza, J. M. and A. van Oudenaarden (2005). "Noise propagation in gene networks." *Science* 307: 1965–1969. DOI: 10.1126/science.1109090 5

Quillardet, P. and M. Hofnung (1993). "The SOS chromotest: a review." *Mutat. Res.* 297(3): 235–279. 3

Quillardet, P., O. Huisman, et al. (1982). "SOS chromotest, a direct assay of induction of an SOS function in *Escherichia coli* K-12 to measure genotoxicity." *Proc. Natl. Acad. Sci. USA* 79(19): 5971–5975. DOI: 10.1073/pnas.79.19.5971 3

Reifferscheid, G. and J. Hell (1996). "Validation of the SOS/*umu* test using test results of 486 chemicals and comparison with the Ames test and carcinogenicity data." *Mutation Research-Genetic Toxicology* 369(3–4): 129-145. DOI: 10.1016/S0165-1218(96)90021-X 3

Ron, E. Z. (2007). "Biosensing environmental pollution." *Curr. Opin. Biotechnol.* 18(3): 252–256. DOI: 10.1016/j.copbio.2007.05.005 5

Sorensen, S. J., M. Burmolle, et al. (2006). "Making bio-sense of toxicity: new developments in whole-cell biosensors." *Curr. Opin. Biotechnol.* 17(1): 11–16. DOI: 10.1016/j.copbio.2005.12.007 5

Tecon, R. and J. R. van der Meer (2006). "Information from single-cell bacterial biosensors: what is it good for?" *Curr. Opin. Biotechnol.* 17(1): 4–10. DOI: 10.1016/j.copbio.2005.11.001 5

Tecon, R. and J. R. van der Meer (2008). "Bacterial biosensors for measuring availability of environmental pollutants." *Sensors* 8: 99–105. DOI: 10.3390/s8074062 5

Ulitzur, S. and I. Goldberg (1977). "Sensitive, rapid, and specific bioassay for the determination of antilipogenic compounds." *Antimicrob Agents Chemother* 12(3): 308–313. 4

Ulitzur, S., T. Lahav, et al. (2002). "A novel and sensitive test for rapid determination of water toxicity." *Environ Toxicol* 17(3): 291–296. DOI: 10.1002/tox.10060 4

van der Meer, J. R. and S. Belkin (2010). "Where microbiology meets microengineering: design and applications of reporter bacteria." *Nat. Rev. Microbiol.* 8(7): 511–522. DOI: 10.1038/nrmicro2392 5

van der Meer, J. R., D. Tropel, et al. (2004). "Illuminating the detection chain of bacterial bioreporters." *Environ. Microbiol.* 6: 1005–1020. DOI: 10.1111/j.1462-2920.2004.00655.x 5

Yagi, K. (2007). "Applications of whole-cell bacterial sensors in biotechnology and environmental science." *Appl. Microbiol. Biotechnol.* 73(6): 1251–1258. DOI: 10.1007/s00253-006-0718-6 5

CHAPTER 2

Genetic Engineering Concepts

2.1 INTRODUCTION TO GENETIC SENSING/-REPORTING CIRCUITS

2.1.1 CENTRAL IDEA

The central idea of bioreporters is to interrogate a biologically relevant pathway or behavior with the help of a reporter protein. Ideally, the reporter protein should be highly sensitive and easy to measure; otherwise, there is no point in replacing the reaction of the original biological pathway with the reporter. The use of reporter proteins is so standard in modern cellular and molecular biology, for example, to study gene expression or to follow protein localization and transport in cells, that one may ask the question why this is relevant to discuss at all. The difference in approach does not come so much from the application of reporter proteins, but from the underlying purpose. In most biological research using reporters, the purpose is not the construction of a biological entity that measures a particular target chemical or condition with the concomitant production of a quantifiable response. Here we are concerned exactly with the question how to design and construct bacterial strains, which can quantify chemical concentrations, fluxes or specific target 'conditions' in their environment via the production of reporter proteins. Not necessarily, but in many cases, one exploits hereto an existing sensory protein or signaling pathway in the cell, and couples the fact that the natural sensory or signaling pathway can 'sense' or 'detect' molecular events to the production of a quantitatively interpretable and easily measurable signal output in the form of a reporter protein (Figure 2.1).

2.1.2 INTERCEPT DESIGN

Research and design in bioreporter construction is basically divided in two conceptual directions. The first of these consist of designs that try to intercept or eavesdrop on an existing signaling pathway in a cell by means of a reporter construction, like the SOS-response type reporters mentioned in Chapter 1. I will elaborate more on those later, but a few examples may help to understand this type of design here (Figure 2.2). Imagine one's purpose is to determine how bacteria react to the presence of a plant root. Because plant root communities are complex ecosystems with multiple species, various types of chemicals and signaling factors, it will not be so easy to find a method to measure the reaction of one particular bacterial species in that system. One method could be to find a key cellular 'node' in the cells of interest and couple reactivity of this node to reporter protein formation. Reporter protein synthesis in cells then approaches the true reaction of the cells, provided that synthesis in itself does not change the behavioral response of the cells [Leveau and Lindow,

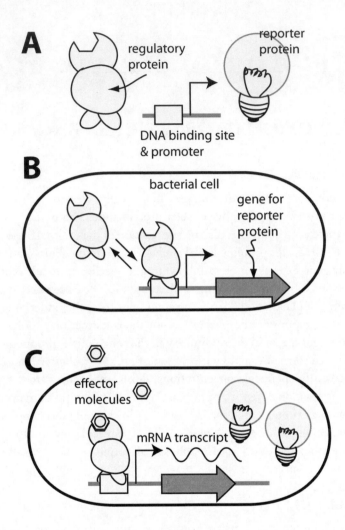

Figure 2.1: Conceptual layout of a bacterial gene reporter (bioreporter). (A) Main components for the functioning of a simple sensor/reporter circuit: a regulatory protein with sensory function, the DNA switch with the DNA binding site of the regulatory protein and the promoter for RNA polymerase, and a reporter system that functions as the final output. (B) The components have to be furnished to the cell in form of DNA, by which the cell can produce the regulatory protein. This can control expression of the reporter protein via its specific binding site near the promoter. (C) In the presence of effector molecules or target conditions for which the system is designed, the regulatory protein will elicit reporter gene expression and the formation or activity of the reporter protein can be measured.

Intercept design

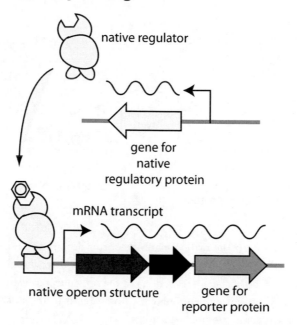

Figure 2.2: Intercept design. In the intercept design, the gene for the reporter protein is placed in the genome of the bacterial reporter cell in such a way as to minimize its impact on the signaling chain or network that it is targeted to analyze. The reporter gene would thus be the only foreign element in such a cell. The goal of the intercept design is to quantify the response of the native cell in its natural setting with minimal impact on the cell itself.

2002]. I would call this layout an interception design. In another experiment, the goal might be to measure the amount of signaling molecules that are released by animal tissue and recognized by a bacterial pathogen. In this case, the design would consist of coupling formation of reporter protein to the activation of the signal recognition pathway in the pathogen. Important for the interception design is thus that one minimizes the impact of the reporter construction on the functioning and physiology of the cell in which it is embedded, to ensure that measurement of the reporter comes as close as possible to the real behavior of the cell (without the reporter).

2.1.3   ORTHOGONAL DESIGN

In the second type of reporter designs, there is no particular interest in the native functioning of the host cell, but this is used simply as a living construction *chassis* for the production and maintenance of the different components of the sensor/reporter circuit (Figure 2.3). This can go so far as to

Orthogonal design

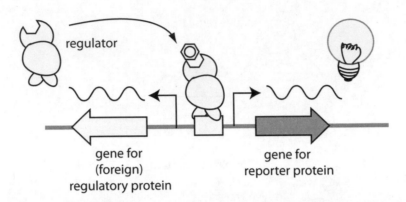

Figure 2.3: Orthogonal design. In the orthogonal bioreporter design, the host cell is simply used as a production chassis for the different components of the sensor/reporter circuit. There is no longer the specific objective of analyzing the host cell's response, but the only purpose is the optimal production and functioning of the circuit. All its components may thus be foreign to the host cell and may have been tailored for specific and optimal functioning.

completely try to uncouple the functioning of the sensing/reporting circuit from all other cellular influence (i.e., *orthogonal*), other than the molecular components necessary for producing the system (e.g., RNA polymerase or ribosomes). New studies have shown that it would even be possible to produce orthogonal translation factories in the cell next to the native ones An and Chin [2009]. The goal here is thus to make a living sensory cell that is optimized according to the requirements needed for target detection. Examples here could include a sensor cell that has an as low as possible method detection limit, that displays the widest linear concentration range of detection of the target compound, or that couples sensing to a completely different type of reaction[3].

Whereas the two different concepts of reporter design may be useful for our understanding, most literature examples obviously do not easily fall in these categories, often because of technical and biological constraints on the experimental system in question, or because the research was not meant for synthetic reporter design. Examples given below will thus unavoidably describe mixtures of several design ideas.

2.1.4   DESIGN PARTS

In its simplest version, the design of a genetic circuit for sensing and reporting starts with the following considerations. First, a choice must be made on the sensory protein or set of proteins that

---

[3]See, for example recent iGEM competitions (http://2009.igem.org/Team:Groningen).

will enable the molecular detection event or that form the signaling chain leading to reporter protein production. A very widely deployed layout makes use of bacterial regulatory proteins that have both a sensory domain (for signal perception) and a regulatory domain (for controlling expression of one or more specific target genes). Typically, the regulatory domain has the capacity of binding to specific regions on the DNA nearby the genes to be regulated and, secondly, to interact in some form with RNA polymerase (Figure 2.1). The protein can thus interact with a chemical or another protein (the sensing event) and relay this sensing to a control of gene expression via its DNA binding capacity and its interaction with RNA polymerase. The second (implicit) element of choice for the sensing/reporting circuit is the DNA region containing the binding site for the regulatory protein and for RNA polymerase (i.e., the promoter). I will call this DNA region the *DNA switch*, to collectively include *activator* or *operator* sites (which bind transcription factors), and to differentiate it from the *promoter*, which is the binding site for RNA polymerase on the DNA. As regulatory proteins may have different DNA binding sites, which may be near to those of undesired auxiliary DNA binding proteins, the choice for the DNA switch is not always trivial. The third element is then the gene for the reporter protein that has to be produced. As will be described more in particular below, numerous reporter proteins have been employed that all can have specific advantages and disadvantages (cf. [Daunert et al., 2000]). For the design of useful sensor/reporter constructions, most researchers have applied reporter proteins such as luciferase or autofluorescent proteins, which are exogenous to the host cell (See Annexes). In this manner, there is no or little interference between recording the reporter protein's presence or activity, and background presence or activity of native host cell proteins.

The actual construction of the bioreporter then consists of combining or placing the DNA fragments for the three different components in the host cell of choice. Obviously, in the interception design, it could be sufficient to place a gene for the reporter protein in transcriptional fusion with the native target promoter or operon (Figure 2.2). In the orthogonal design, one needs to combine the gene for the regulator protein of choice, the DNA switch with the binding site for the regulator protein(s) and the promoter, which have to be fused to the gene for the reporter protein. Proper expression of the gene for the regulator needs to be ensured and all DNA fragments need to be adequately assembled together and placed in the genome of the host organism (Figure 2.3). Standard genetic techniques can be used for isolation and production of the desired DNA fragments, for example, by polymerase chain reaction (PCR) amplification using genomic DNA of organisms carrying the required genes for regulatory proteins or reporter proteins, and the DNA binding sites and promoter sequences. Alternatively, one can order the DNA sequences in the requested design by chemical DNA synthesis. More and more DNA 'parts' are deposited in virtual repositories such as the Biobricks Foundation, which list pertinent information and DNA sequence of the parts, plus user information if available[4].

Finally, as will be explained in more detail below, the design and functioning of the sensor/reporter circuit or its components is dependent on various technical and biological constraints.

---

[4]http://parts.mit.edu/registry.

Such constraints can include: reducing background expression from the reporter gene, increasing or decreasing sensitivity of response, or changing the specificity of the detection reaction. More complex layouts may include multi-component sensory and signaling chains, multiple regulatory circuits in series or multiple reporter elements in parallel.

2.2 USE OF TRANSCRIPTIONAL ACTIVATORS

2.2.1 CHOICE OF REGULATORY PROTEINS

Inasmuch as the specificity of detection in the sensor/reporter circuit is dependent on the recognition specificity of the sensory protein, it is important to pay sufficient attention to the choice of sensory protein. As explained above, for the development of interception-type bioreporters, one may be more easily limited by the nature of the sensory protein because it is exactly its biological nature that one wishes to investigate. For orthogonal-type bioreporters, one can try to select the most appropriate sensory protein for the target chemical detection or target condition one is interested in. In this case, one could try to exploit all the existing biological variability for a given effector-sensor protein interaction or use molecular and modeling tools to optimize such interactions.

One important class of sensory proteins in bioreporter construction has traditionally been bacterial multi-domain transcription activators (Figure 2.4). There have been a number of reasons for the choice of bacterial transcription activators as sensing and, as explained above, signal relay elements in sensor/reporter circuits. First of all, detailed knowledge exists on protein structure, activation mechanisms and DNA binding characteristics of a number of different classes of bacterial transcription activators [Diaz and Prieto, 2000, Tropel and van der Meer, 2004]. However, it is fair to say that despite many efforts, there is still only a limited picture on the activation mechanisms exerted by chemical effectors on such proteins. Secondly, a number of them were relatively easy to manipulate, seemed to have relatively clear chemical effectors and could thus be tested for proof of principle in biosensing. Third, because of the clear initial focus to have bioreporters for measurement of toxic pollutants, rather than say, of glucose, many studies in literature have used bacterial transcription activators implicated in pollutant recognition for the cell.

Transcription activators – as their name implies – increase the rate of RNA polymerase transcription from otherwise poorly transcribed promoters, usually upon a specific triggering of the activator protein by another protein (e.g., phosphorylation signaling cascade) or by direct interaction with a chemical *effector*. In the absence of the transcription activator or of the triggering, such promoters are mostly, but not completely, silent. The fact that most promoters are not completely silent even in the absence of effector is the main reason for the occurrence of background expression of the reporter protein in the case such system is applied for sensing/reporting.

A primary and central question in bioreporter design is thus how to find the proper transcription activator for the target chemical or condition of choice. In early designs, one would typically assume that growth of a bacterium on a particular chemical compound would mean that the cell has a way of recognizing the compound and adjusting synthesis of the metabolic enzymes needed for its degradation. As an example, in order to make a bioreporter for naphthalene, one would take

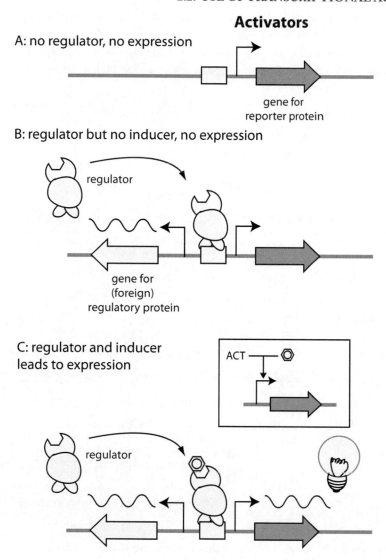

Figure 2.4: Activators. Transcription activators are characterized by the fact that, (A) in their absence, there is often no expression of the promoter, (B) in their presence but in absence of the chemical effector, there is no expression of the target promoter, whereas (C) gene expression is only elicited in the presence of the inducer. Transcription activators can thus prevent RNA polymerase to start at certain promoters and stimulate RNA polymerase in the presence of effectors (or otherwise signals that will activate the regulatory protein). Typically, the activation process requires a conformational change of the regulatory protein and direct interaction with RNA polymerase. As outlined in the design, transcription activators often also control their own gene expression on DNA switches containing divergent promoters.

a bacterial strain degrading naphthalene, in order to exploit the molecular components for 'naphthalene' recognition. One of these is *Pseudomonas putida* strain pPG7 carrying the NAH plasmid, on which the *nah* operons are located that encode the full degradation of naphthalene (Figure 2.5). The central regulatory element is a transcription activator called NahR, which stimulates expression

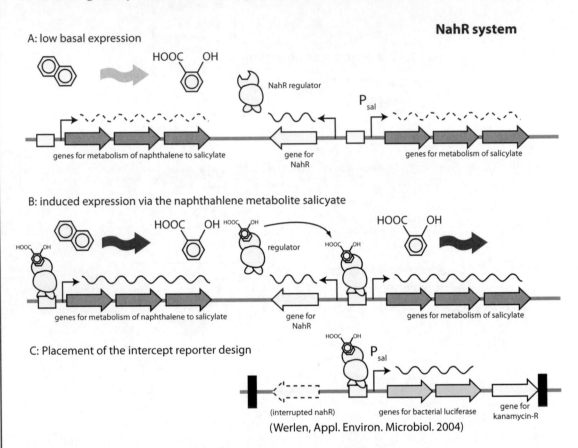

Figure 2.5: The NahR regulatory system. NahR is the central transcription activator in controlling expression of the genes necessary for naphthalene metabolism in the bacterium *Pseudomonas putida*. (A) Complication of the NahR system is that naphthalene itself is not the effector leading to NahR-mediated activation and in the absence of naphthalene metabolism transcription from the *nah* genes is very low. (B) When naphthalene is coming into the cells it is metabolized to salicylate, due to basally expressed metabolic enzymes. Once salicylate is formed this triggers further and higher expression of the system. (C) NahR and the DNA switch named P_{sal} have been employed for the construction of naphthalene sensing bioreporter strains. In this drawing, an interception design was chosen, which was placed within the boundaries of a mini-transposon on the chromosome of *P. putida*. The *nah* genes are present on the NAH7 plasmid.

of the *nah* genes from two promoters, one in front of *nahG* and one in front of *nahAa* [Schell, S., 1993, Schell and Poser, 1989]. Complication to this system is that NahR is not directly responsive to naphthalene(s), but to salicylate, which is a metabolic intermediate of naphthalene degradation. In the native biological configuration, the cell transcribes the *nah* genes at a low rate and produces a small amount of metabolic enzymes even in the absence of naphthalene. Once it encounters naphthalene, this is converted and salicylate is formed, which triggers further expression of the naphthalene metabolic enzymes. For the design of a naphthalene bioreporter based on *P. putida* and its NahR regulator, one must therefore include the genes to generate intracellular salicylate from naphthalene in order to obtain a reporter signal from the cells to externally added naphthalene. One way of achieving this was to utilize the original *P. putida* strain with the NAH7 plasmid and introduce a second (chromosomal) copy of the *nahR* gene plus the original DNA switch containing the NahR binding site and the promoter (P_{sal}) fused to a reporter gene (i.e., interception design) [Werlen et al., 2004]. An alternative way was to introduce both a NahR-regulated gene reporter circuit and the set of genes needed for naphthalene to salicylate metabolism into a new host cell (e.g., *Pseudomonas fluorescens*) [King et al., 1990].

Along similar lines, most of the bacteria that were well known at the time for their capacity to degrade organic pollutants and of which the regulatory systems had been characterized have been applied for the construction of bioreporters. In most of these, the central sensing element was a transcription activator, but not necessarily of the same protein family [Tropel and van der Meer, 2004]. As an example, the aforementioned NahR protein belongs to the LysR family of transcription activators of which also various other members have been applied for sensory purposes (e.g., ClcR, TfdR, CbnR or DntR, see Annexes). As a matter of fact, all those LysR transcription activators respond to metabolites, which have to be generated intracellularly. This complicates their design because it has to be anticipated that a target compound needs to be metabolized before being detected by the sensor/reporter circuit. For example, the regulators ClcR, TfdR, CbnR all are activated by *cis,cis*-chloromuconates which arise during metabolism of chlorocatechols. Apart from not being commercially available compounds, externally added chloromuconates are very poorly taken up by the reporter cells. Should one wants to deploy those sensor/regulators for reporter circuits one has to ensure that chloromuconates are formed. Typically, chloromuconates can be formed from chlorocatechols, which themselves arise in metabolism of a variety of chloroaromatic compounds [Diaz, D., 2004, Reineke and Knackmuss, 1988]. One could thus imagine using such sensor/reporter circuits to detect metabolic network 'nodes', in which more than one target compound can be sensed (Figure 2.6).

2.2.2   THE XYLR/DMPR FAMILY OF TRANSCRIPTION ACTIVATORS

A further important class of sensory proteins stems from the XylR/DmpR family of transcription activators, which is a subfamily of the NtrC-type regulators [Galvao and de Lorenzo, 2006]. Members of the XylR/DmpR subfamily regulate a variety of aromatic degradation pathways, such as those for *m-*,*p*-xylene and toluene (e.g., XylR of *P. putida* carrying the TOL plasmid) [Ramos et al., 1997],

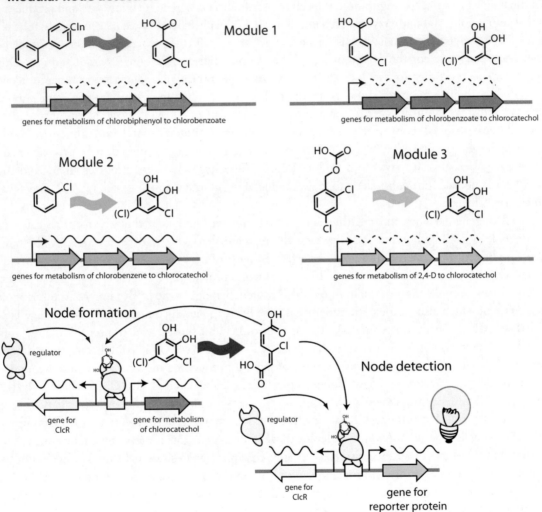

Figure 2.6: Node design. Multiple metabolic reactions may give rise to the same metabolite (here, chloromuconate). By having the sensor/reporter construct detecting this metabolite, one could assay multiple different target compounds, provided the reporter cell carries the gene modules for producing the metabolite from the target compound. Example of node design using *tfdR*, see [Hay et al., 2000].

for (methylated) phenol (e.g., DmpR of *Pseudomonas* sp. CF600) [Shingler and Moore, 1994], for *o*-xylene (TouR of *P. putida*) [Arenghi et al., 1999], for xylenes, ethylbenzenes and toluene (TbuT of *Ralstonia picketti*) [Leahy et al., 1997], for phenanthrene and naphthalene (PhnR of *Burkholderia sartisoli*) [Laurie and Lloyd-Jones, 1999], or for 2-hydroxybiphenyl (HbpR of *P. azelaica*) [Jaspers et al., 2000]. The group of XylR/DmpR transcription activators has the advantage that the chemical effector eliciting the activation process in most cases is also the primary substrate of the degradation pathway [Tropel and van der Meer, 2004]. This means that no further cellular metabolism is needed, like for NahR and naphthalene, to detect the target chemical. On the other hand, many XylR/DmpR members require additional cellular proteins for optimal functioning (e.g., Integration Host Factor) [Perez-Martin and de Lorenzo, 1995], and their cognate promoters may be controlled by other auxiliary cellular regulatory systems (e.g., PprA [Vitale et al., 2008]). This can complicate their application in orthogonal bioreporter constructions. Two examples will be explained here in more detail, that of XylR and of HbpR (Figures 2.7 and 2.8), which are exemplary for a number of engineering reasons.

A large number of engineering efforts have concentrated around the XylR system of *P. putida* mt-2. The XylR system originally occurs on the TOL plasmid pWW0 of *P. putida* where it controls expression of the *xyl* genes involved in degradation of toluene, *m*- and *p*-xylene [Ramos et al., 1997]. The XylR activator protein hereto acts at two DNA switches, called P_S and P_U. Regulation by XylR has been the subject of numerous detailed studies because of its exemplary character for a large protein family and because of its complexity. In its native form, XylR is one of the master regulators for expression of the *xyl* catabolic genes on the plasmid pWW0 from *P. putida* mt-2 that collectively endow the cell to degrade toluene, *m*- and *p*-xylene. XylR drives expression of two promoters in the *xyl* gene clusters: the P_S-promoter in front of *xylS* and the P_U-promoter in front of *xylU*. XylS, on its turn, is also a regulatory protein that is needed for further expression of another part of the *xyl* genes from a promoter called P_M [Ramos et al., 1997].

Typical for proteins like XylR is that they regulate expression at a longer distance from the promoter, and, secondly, that the promoter is not σ^{70} but σ^{54}-dependent [Perez-Martin and de Lorenzo, 1995, Shingler, V., 1996]. In contrast to σ^{70} promoters that have RNA polymerase binding site motifs at positions -10 and -35 relative to the transcription start site, promoters that are σ^{54}-dependent typically have a -12, -24 motif [Shingler, V., 1996]. The binding sites for XylR (so-called *Upstream Activation Sequences*, or UAS) are located between 140 and 180 bp upstream of the transcription start, and they are thought to encompass two regions of 16-bp each with imperfect inverted repeats (Figure 2.7) [de Lorenzo et al., 1991]. XylR is supposed to form a larger multimeric structure on the DNA, not unlikely a hexamer, which induces a strong bending of the DNA [Garmendia and de Lorenzo, 2000]. A further bending in the control region is needed, probably to bring XylR in optimal contact with σ^{54}-RNA polymerase [Perez-Martin et al., 1994]. This bending is facilitated by Integration Host Factor, for which one or more binding sites are located in the same control region (Figure 2.7) [Perez-Martin and de Lorenzo, 1995]. The control region of P_S thus actually contains two promoters, one for *xylR* itself and the one for *xylS*, which

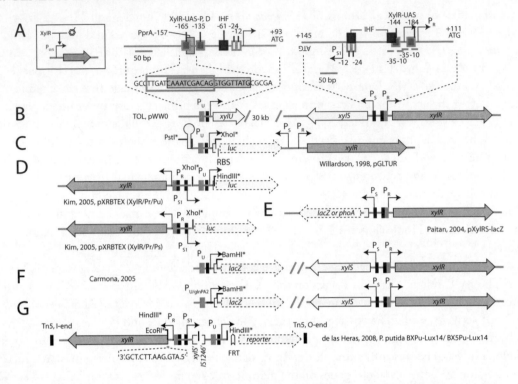

Figure 2.7: XylR designs. (A) and (B): natural configuration of the *xyl* genes and the major XylR and XylS regulated promoters. The binding sites for XylR on P_U are located some 150 bp upstream of the transcription start and are characterized by two sequence regions of high similarity (open box in Figure 2.7 (A)). Further control elements are an IHF binding site (as a black box) and a binding site for the protein PprA, which overlaps with the XylR Upstream Activating sequences (grey shaded box). XylR controls a σ^{54} promoter, which is bound by the alternative sigma factor σ^{54}-RNA polymerase complex at a -12, -24 sequence element. The configuration of the P_S switch is similar, but different in detail. XylR binding sites are again found at a distance of some 150 bp to the *xylS* transcription start, but these overlap with the two regular σ^{70}-dependent promoters (hence -10, -35 boxes) controlling *xylR* expression itself. IHF binding sites are present in the switch, but not one for PprA. Not unimportantly, a second σ^{70}-dependent low constitutive promoter is also controlling basal expression of *xylS*. XylR is expressed at a more or less low constitutive rate, whereas both P_U and P_S are essentially silent in absence of XylR and/or its cognate effectors. (C) Willardson design, to place *xylR* on the same construct as the P_U-*luc* reporter gene fusion. (D) Kim design, in which *xylR* is placed divergently of the P_U-driven reporter gene. Note how the direction of the constitutive P_{S1} promoter may cause high background of the reporter gene. (E) Paitan design, using the XylR-regulated P_S promoter to drive reporter gene expression. (F) The split design of Carmona and coworkers, providing *xylR* in trans elsewhere on the genome of *P. putida* and using hybrid P_U promoters devoid of exponential phase silencing control. (G) The cassette design presented in de las Heras et al. [2008], in which the reporter gene is placed within a mini-Tn5, in which subsequently different regulatory genes and promoters can be docked.

is regulated by XylR. Expression of *xylR* is controlled by two σ^{70}-promoters, the sites of which overlap with the XylR binding site and with an IHF binding site. XylR thus exerts a slight autoregulatory control on itself. It should be noted that a second σ^{70}-promoter, called P_{S2}, is providing a low rate of transcription of *xylS*, even in the absence of XylR-mediated control [Marques et al., 1998, Marques and Ramos, 1993]. As can be seen from the various designs of XylR sensor/reporter elements, the presence of the *xylS* P_{S2}-promoter can result in enhanced undesired background expression of the reporter element.

The second XylR-controlled promoter is P_U, a region of some 300 bp upstream of the *xylU* gene. This switch again displays the elements described above for optimal XylR regulation, i.e., a σ^{54}-RNA polymerase binding site, an IHF binding site and UASs for XylR interaction (Figure 2.7) [Ramos et al., 1997]. Puzzling for a long time, however, has been a silencing of P_U expression during certain physiological conditions both in *E. coli* and *P. putida*, which had been called 'exponential phase silencing' [de Lorenzo et al., 1993]. Recently, it was discovered that a binding site for a protein named PprA is overlapping with the UASs of XylR on P_U (but not P_S) [Vitale et al., 2008]. This protein is in competition with XylR for its binding and seems to be responsible to inhibit P_U expression under conditions of excessive other carbon sources than toluene. Because this region is present in all constructions so far that have used the P_U promoter, the PprA protein can exert a negative influence on the reporter output.

In the original configuration, the gene *xylR* is located opposite of the gene *xylS* (Figure 2.7), whereas P_U finds itself in front of the gene *xylU* at a distance of some 30 kb to the *xylR* gene. Both switches are characterized by a high complexity of elements controlling its functioning. Various bioreporter constructs have employed the P_U promoter, in which case the gene for *xylR* has to be provided *in trans* (Figure 2.7). When working in host strains outside the original *P. putida* with the TOL plasmid, the reporter construct typically has to couple the *xylR* gene and the P_U-switch on one vehicle, whereas they do not occur as such in their native configuration. For example, in the early Willardson design, a fragment containing the *xylR* gene (under control of the P_R – P_S switch) is connected downstream of the P_U-*luc* reporter [Willardson et al., 1998]. Readthrough into P_U is protected by use of a transcription terminator, for which two appropriate restriction sites were created by PCR design. Even though P_U-*luc* is a transcriptional fusion, a small fragment of the *xylU* gene is included in the design, but no clear information is presented on the sequence connection between *xylU'* and *luc*, except that *luc* carries its own Shine Dalgarno sequence for ribosome binding [Willardson et al., 1998]. The complete construction was placed on a commercially available *luc* vector pGL2 in *E. coli*. In a second design, they removed P_U completely and placed the reporter gene directly under control of P_S (and perhaps its low constitutive promoter, Figure 2.7 (C)) [Willardson et al., 1998]. In the Kim design of 2005 [Kim et al., 2005], the *xylR* gene is placed upstream and divergently of the reporter gene. In one of their configurations, which the authors say has a high background, both P_S and P_U switches are in the same transcriptional direction. It is likely that the high background in this design results from the second low constitutive promoter that drives *xylS* expression [Marques et al., 1998]. Restriction sites were engineered for cloning, but, otherwise, the switch sequences were amplified by PCR without introducing further changes. In this design, only a few codons of *xylU* remain in front of the reporter gene. In the Paitan design of 2004 [Paitan et al., 2004], P_S is directly used to drive reporter gene expression and a gene fragment starting within *xylS* and encompassing the full length *xylR* is cloned in front of the reporter gene.

In the designs by Carmona and coworkers, of which only two are reproduced here, again P_U is used to drive reporter gene expression, but *xylR* is placed under its native control elsewhere on the chromosome of *P. putida* [Carmona et al., 2005]. The interesting point in these designs is that the authors produced hybrid promoters, in which the IHF and σ^{54}-RNA polymerase binding sites of the P_U switch were replaced by those of other σ^{54}-dependent switches, such as *glnA2p* and *nifH*. This was done in an attempt to reduce the transcriptional silencing effect on P_U during exponential growth on complex medium. Indeed, the P_U-*glnA2p* hybrid was less subject to the silencing effect [Carmona et al., 2005]. More recently, the group around Victor de Lorenzo produced new synthetic designs in which *xylR* mutants can be used and integrated into the genetic circuit by cassette exchange [de las Heras et al., 2008]. The reporter gene is again expressed from native P_U but without traces of *xylU*. As a result of the cassette exchange technique the FRT yeast flippase sequence recombination element is left upstream of P_U. This, the authors note, decreases background expression from P_U [de las Heras et al., 2008]. Note that traces of *xylS* and an IS element upstream of P_U are remaining in the construct (Figure 2.7 (G)). Despite all the nice cloning strategies, the XylR-P_U or P_S-systems are not ideal from a practical point of view because of the relatively high method of detection limit for compounds such as xylene or toluene (lower mM range, see Annexes).

2.2.3   THE HBPR SYSTEM

The *hbp* system from *Pseudomonas azelaica* operates in a similar way as the XylR-mediated control, but with some important differences (Figure 2.8). HbpR controls expression of two promoters in a short cluster of three *hbp* genes, in front of *hbpC* and *hbpD* [Jaspers et al., 2001a, 2000, 2001c]. The control region containing the P_c promoter (in front of *hbpC*) shows a similar order of control elements as described for XylR in the case of the P_s promoter. The P_c promoter contains the typical -12, -24 boxes representative for σ^{54}-RNA polymerase. An IHF binding motif is located at a distance of some 80 bp to the transcription start site, and IHF indeed was implicated in optimal P_c expression [Jaspers et al., 2001c]. A bit further away as in the XylR system is the binding region for HbpR with the typical double UAS and their inverted repeats, although HbpR was actually shown to cover a region extending to both ends of the UAS [Tropel and van der Meer, 2002]. Divergently oriented, but not overlapping with the UASs for P_c, is the -10, -35 σ^{70}-dependent promoter that drives expression of *hbpR* itself. Most curiously, however, and probably a consequence of natural recombination mechanisms is a short fragment of a gene similar to *hbpR*, which is located at some 100 bp downstream of the *hbpR* transcription start site [Jaspers et al., 2001c]. This gene fragment is then again followed by a second set of UASs to which HbpR can also bind (called UAS-C4

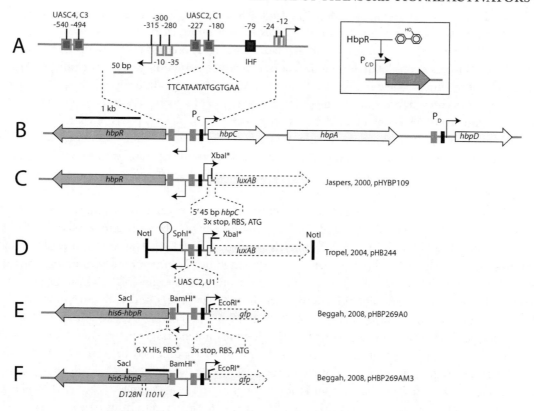

Figure 2.8: The HbpR regulatory system of *P. azelaica* and its applications in various sensor/reporter circuits. In its native form (A, B) HbpR regulates expression from two promoters, in front of *hbpC* (called P_C) and in front of *hbpD* (called P_D). Most designs employed the P_C switch. Similar as for XylR, the HbpR protein activates RNA polymerase with the σ^{54} factor at the -12, -24 sequences of the promoter. The main HbpR binding site (called UAS C-1, C-2) is located between 170 and 240 bp upstream of the transcription start, however, in the native configuration there is a second HbpR binding site at a distance of around -500 (called UAS C-3, C-4). Binding sites for HbpR upstream of *hbpD* are called UAS D-1, D-2. Binding sites for HbpR are not identical to those of XylR, but still highly similar. Again, an IHF binding site in the P_C switch aids for optimal expression. The *hbpR* gene itself is transcribed from a σ^{70}-dependent promoter, which is not overlapping with UAS C-1, C-2, but upstream of UAS C-3, C-4. (C) The original configuration of the Jaspers design fuses the *luxAB* luciferase genes directly downstream of a small *hbpC* fragment but avoids translational coupling via the use of stop codons. (D) In *trans* designs by Tropel et al to test wild-type and hybrid P_C-P_U promoters. (E) and (F) Reconstructed plasmid designs by Beggah et al, facilitating A-domain gene fragment exchanges to test mutant effector profiles. The constructs have slightly altered *hbpR* starts and do not further include a *hbpC* fragment in front of the *egfp* reporter gene. Boxes and arrows point to DNA motifs important for functioning of the *hbpR*-P_C system.

and C3) [Tropel and van der Meer, 2005]. These UASs are not necessary for optimal functioning of the P_c promoter, but exert some control on *hbpR* expression itself. Further contrast to the XylR-controlled P_U promoter is that no exponential phase silencing is observed for P_c, neither in *P. azelaica* nor in *E. coli* [Jaspers et al., 2000]. Different reporter circuit constructions that have employed HbpR and the P_c switch are outlined in Figure 2.8. In the Jaspers design of 2000 [Jaspers et al., 2000], the native *hbpR*-P_C configuration was retained for construction of a luciferase reporter circuit. In this case, 15 codons from *hbpC* are retained, but no translational coupling is possible via the placement of stop codons in three frames downstream of the *hbpC* fragment. Optimal luciferase translation can occur via its own ribosome binding site properly spaced from the translation start. This circuit was exploited successfully both on plasmid in the heterologous host *E. coli* and on mini-transposon in the chromosome of *P. azelaica* [Jaspers et al., 2000]. The system was also used for the design of a hybrid switch, with the idea of obtaining activation by two regulatory systems simultaneously in the same cell. This could be accomplished by creating a hybrid binding site consisting of UAS C-2 for HbpR and UAS U-1 from P_U for XylR [Tropel et al., 2004]. Transcription read-through from upstream of the hybrid P_C-P_U switch was prevented by placement of a terminator structure. The hybrid circuit was used either on plasmid or on mini-transposon in the chromosome of *P. azelaica* and *P. putida*, and induction by both *m*-xylene (via XylR) and 2-hydroxybiphenyl (via HbpR) was demonstrated [Tropel et al., 2004]. In the 2008 designs by Beggah and coworkers, an easy exchange module was created to replace the A-domain gene fragment in wild-type *hbpR* for mutant A-domains [Beggah et al., 2008]. This design also re-engineered the start of *hbpR* with a sixfold histidine tag, while leaving the other elements of the P_C switch intact. Only the first codon of *hbpC* was left in this design whereas translational coupling to the *egfp* gene was avoided by placing stop codons in three reading frames (Figure 2.8). The *hbpR*- P_C system in its various designs was applied successfully for a variety of bioreporter assays (see Chapter 3).

2.3 USE OF TRANSCRIPTIONAL REPRESSORS

2.3.1 REGULATORS FROM HEAVY METAL RESISTANCE

Transcriptional repressors have also been applied numerous times as the control element for sensor/reporter circuits in bacteria. Similar to transcription activators, repressors keep the transcription rate from their target promoters very low under non-inducing conditions, but they do so typically by preventing access of RNA polymerase to its promoter. The DNA binding sites for repressors (called *operator sites*) are often located directly in overlap with or downstream of a promoter, in which case physical presence of the repressor protein on the DNA hinders RNA polymerase in transcription. In case of effector recognition the repressor protein changes its affinity for its operator sequence, the net effect of which is that per unit of time it will less occupy the DNA, which allows RNA polymerase to start transcription more often [Lloyd et al., 2001]. In contrast to transcription activators, therefore, repressors have no need to activate RNA polymerase to start transcription from unproductive promoters in the presence of a signal, but they lose their affinity for the DNA and release their repression (Figure 2.9).

Repressors

A: no regulator, expression

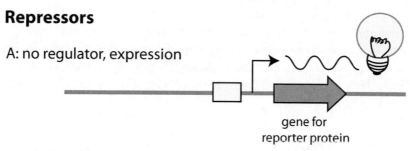

gene for
reporter protein

B: regulator but no effector, no expression (repression)

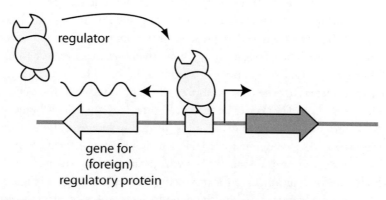

regulator

gene for
(foreign)
regulatory protein

C: regulator and effector leads to expression (derepression)

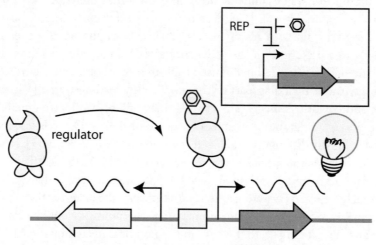

regulator

Figure 2.9: Transcriptional repressor functioning.

The use of transcriptional repressors for bioreporter designs most often coincided with target-ing a completely different spectrum of chemicals or conditions as described above for transcription activators. For example, in contrast to transcription activators that are most often implicated in con-trol of metabolism of organic compounds, transcriptional repressors are often in control of metal resistance. Consequently, bioreporters to detect (heavy) metals rely on transcription repressors. Two of the most widely deployed heavy metal sensing systems are those for mercury and for arsenic, and use the MerR and ArsR repressors, respectively.

2.3.2   THE MERR SYSTEM

The genes for mercury resistance form part of several transposable elements such as Tn501 or Tn21, which are often found in conjunction with conjugative plasmids [Park et al., 1992]. On all mercury resistance regulons, the gene for MerR is encoded in the opposite direction as the genes producing the resistance mechanism. The MerR protein is a relatively small protein of 145 amino acids, which binds as a dimer on the DNA, spanning a region of 17 bp. The control region is formed by a 72-bp intergenic sequence between *merR* and *merT* that contains both overlapping classical σ^{70} promoters [Summers, S., 1992]. As can be seen in Figure 2.10, the MerR binding site overlaps with the transcription start site for its own gene and is located in between the -10 and -35 boxes of the promoter for the *merT*. As a consequence, MerR acts as a repressor for its own synthesis but, at the same time, as an activator for expression of the *merT* promoter [Summers, S., 1992]. This promoter is the one, which is most typically used for producing the reporter signal (Figure 2.10).

In the Mer sensor/reporter design, the gene for the repressor protein is most often used in its 'native' configuration, that is, under control of its own promoter(s). In the first mercury bacterial sensor/-reporter, Selifonova et al. [1993] cloned a *Bam*HI-EcoRI fragment, which encompassed the full *merR* gene of Tn21 in *E. coli* plus a small part of the *merT* gene, and they fused this transcriptionally to a cassette with the genes for bacterial luciferase [Selifonova et al., 1993]. Virta et al. followed up on the same design but used PCR to incorporate a shorter *merR* fragment plus the control region, without including part of the *merT* open reading frame [Virta et al., 1995]. More recently, Pepi and coworkers used again a similar idea of a restriction fragment for construction of their sensor/reporter circuit, but this time, the source of their DNA was a *P. stutzeri* plasmid pPB that contained a slightly differently organized *mer* operon [Pepi et al., 2005]. In an attempt to detect not only Hg^{2+}, which is the species that is recognized directly by the MerR dimers, but in addition organomercurial compounds, Ivask et al. in 2001 published a design that included a gene for the organomercurial lyase *merB* [Ivask et al., 2001]. In this case, the origin of the *mer* genes was a plasmid pDU1358 from *Serratia marcescens*. Organomercurial compounds would be cleaved by MerB in the reporter cells and liberate Hg^{2+}, that would then again be detected by the MerR sensor. All of the mercury sensor/reporter circuits work quite well in bioassays, achieving in some cases extremely low method of detection limits in the picomolar to low nanomolar range (See Annexes). None of the groups working in this area really made any new 'synthetic' design; all applied the control region and MerR as is.

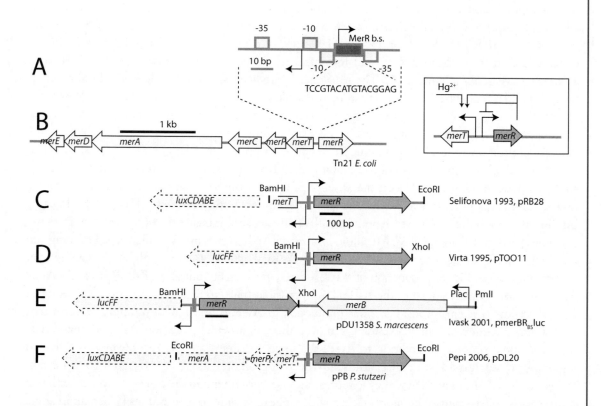

Figure 2.10: MerR repressor designs. The MerR regulatory protein is widely used for the design of bioreporters targeting mercury (Hg^{2+}) and originates in the bacterial defense system to mercury. Different natural varieties of the *mer* system have been applied, such as from Tn*21* in *E. coli*, the plasmid pDU1358 from *Serratia marcescens* or the plasmid pPB from *Pseudomonas stutzeri*. In its natural configuration, MerR regulates transcription from the *merT* promoter and from its own promoter, which are divergently oriented (B). This relatively small intergenic region (the *mer* switch) contains two overlapping σ^{70}-dependent promoters (A), with which the MerR binding site overlaps. MerR thus represses its own transcription by steric hindrance, but it is capable of activating the *merT* promoter when it is detecting Hg^{2+}-ions. In all of the designs (C-F) the *merR* gene is used under the expression of its own promoter on the *mer* switch, and the *merT* promoter is used to drive expression of the reporter gene. The natural genetic circuit is so efficient that mercury can be detected at pico- to nanomolar concentrations. In only one design, a further extrapolation of the spectrum of detectable mercury compouds was achieved by including the gene for organomercurial lyase *merB*.

2.3.3 ARSR-BASED DESIGNS FOR ARSENIC DETECTION

Sensor/reporter constructions for arsenic detection have exclusively been based on the ArsR protein, which is the transcriptional repressor of the arsenic and antimonite resistance mechanism in many bacteria [Rosen, R., 1995, Wu and Rosen, 1993]. Very often, research groups have used the *ars* operon derived from plasmid R773 and R46 of *E. coli*, in which case the *arsR* gene is the first in a series of five genes (Figure 2.11). The resistance is produced by an ATP-dependent efflux pump (ArsBC) and a reductase (ArsA) that can reduce arsenate (As[V]) to arsenite (As[III]) [Rosen, R., 1999]. The efflux system is only pumping out arsenite, but not arsenate. In its native configuration, expression of the five genes is regulated by a control region in front of *arsR*, which contains a σ^{70}-dependent promoter and a 23-bp binding site for an ArsR homodimer. By contrast to the *mer* control switch, the ArsR binding site is not overlapping directly with the RNA polymerase binding site but slightly upstream (Figure 2.11). In the absence of arsenic, the *ars* genes are transcribed at a very low rate, which is necessary to produce ArsR. Upon contact with arsenite (As[III]), ArsR loses its binding affinity to the DNA, which permits RNA polymerase to increase transcription rates from the *ars* genes [Wu and Rosen, 1993]. Metalloid binding to ArsR occurs in a so-called type I binding site in which three cysteines coordinate the metalloid [Ye et al., 2005]. Interestingly, ArsR derepression results not only in higher production of the resistance mechanism, but also of ArsR itself. As ArsR has a very low affinity constant for As[III], it is possible that ArsR is a 'scavenger' for arsenite in the cell at low concentrations. Constant production of ArsR from its own promoter ensures that the system is turned off when arsenite is removed the cell, in which case non-As[III] bound ArsR represses the promoter. In addition to ArsR, the natural system in some cases expresses a second regulator protein (ArsD) with lower affinity to the *arsR* promoter and to As[III]. At high expression levels, ArsD will thus have a tendency to repress the *arsR* promoter and the complete system with two repressors is able to control homeostasis in a wide range of arsenic concentrations [Chen and Rosen, 1997]. In addition to responding to arsenite, the ArsR defense system also reacts to antimonite (Sb[III]) with about the same sensitivity. The *ars* control switch has been applied numerous times for the construction of bioreporter bacteria, but exclusively in its native form with *arsR* as first gene under the control of its own promoter, followed by the reporter gene(s). Bioreporters that employ the native configuration of the *ars* control system suffer from high background signals, which are the result of the low transcription rate that is necessary to produce ArsR and the read-through in the reporter gene. We will see below which strategies have been followed to control the background expression of the reporter gene.

In the simplest designs such as by Tauriainen et al. [1997] and by Scott et al. [1997] the *arsR* gene is under control of the native *ars* promoter, and the gene for the reporter protein is placed downstream of *arsR*. In the Scott design, the *lacZ* reporter is fused translationally to a small fragment of *arsD* [Scott et al., 1997]. Given that they observed very high background signals in such ArsR-reporter constructions, Stocker et al. designed a system to potentially reduce read-through by RNA polymerase from *arsR* into the reporter gene, but maintaining the feed-forward loop of *arsR* expression itself [Stocker et al., 2003]. This was essentially accomplished by placing a small DNA

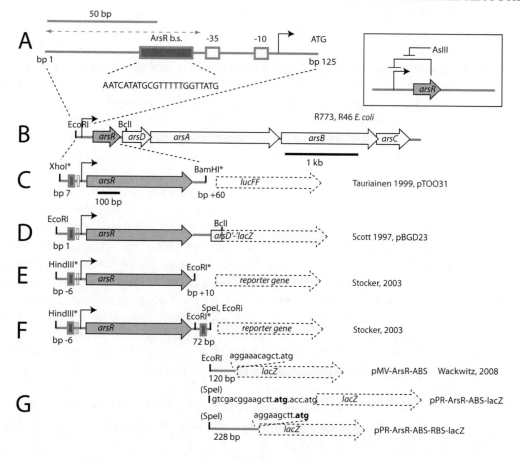

Figure 2.11: ArsR designs. ArsR is the major regulator of the defense system to arsenicals in bacteria. The *arsR* gene is typically the first gene in the defense operon, which produces an arsenite specific efflux pump (ArsBC) and an arsenate reductase (ArsA). The system has the particularity of a feed-forward loop – ArsR represses its own expression and that of the rest of the *ars* operon, but a low transcription rate is necessary to produce the ArsR repressor. ArsR loses affinity for its DNA binding site when it binds to arsenite (As[III]), which causes derepression of the system. The binding site for ArsR is slightly upstream of the promoter (A). In the simplest designs (C, D), the *arsR* gene is used under control of the native *ars* switch and the gene for the reporter protein is placed downstream of *arsR*. This results in a considerable background because of the feed-forward nature of the genetic circuit. (E-G) In the Stocker designs of 2003, the background expression from *arsR* into the reporter gene is reduced by placing a second ArsR binding site (stippled region in panel A) downstream of arsR and upstream of the reporter gene. Further finetuning of the reporter output was achieved in the Wackwitz designs in 2008, who placed various upstream regions for lacZ that either increased or decreased the translation rate, and thereby the sensitivity of arsenite detection by the system. The same thing was achieved by using a set of reporter gene mutants with different catalytic activity.

fragment for a second ArsR binding downstream of *arsR* but upstream of the reporter gene, the idea being that further binding of an ArsR dimer would block RNA polymerase from reading through. Background expression of *gfp* and *lacZ* was indeed largely reduced in *E. coli* carrying such reporter constructs on plasmids, which enabled systematically lower method of detection limits [Stocker et al., 2003]. Further fine-tuning of the reporter output was achieved in the Wackwitz designs in 2008, who introduced various upstream regions for *lacZ* that either increased or decreased the translation rate, and thereby the sensitivity of arsenite detection by the system [Wackwitz et al., 2008]. By changing the sensitivity of detection these authors then proposed bioreporter assays with multiple strains simultaneously, which would alleviate the necessity for external calibration ('traffic light system' – see Chapter 3).

Other bacterial resistance systems have been employed for the construction of heavy metal responsive bioreporters (Annexes). Most notable are the various gene clusters that promote resistance to copper, zinc, cadmium and chromium, originating from the bacterial strain *Cupriavidus metallidurens* CH34 [Collard et al., 1994]. A great deal of biochemical information has also become available on the CadC protein from *Staphylococcus aureus* that regulates expression of the metal resistance mechanism and is responsive to Pb[II], Cd[II], Zn[II], Sb[II] and Sn[II] [Corbisier et al., 1993, Li et al., 2008, Ye et al., 2005]. CadC belongs to the same protein family as ArsR, now called the ArsR/SmtB family, and like ArsR, forms a homodimer that represses transcription of its own gene in a small operon formed by *cadC* and *cadA*, coding for a P-type metal translocating ATPase. Interestingly, CadC has both type I (like ArsR, involving Cys coordination) and type II metal binding sites, which is formed by ligands from Asp, His and Glu residues on both dimers [Ye et al., 2005]. As a consequence, CadC can bind both Cd^{2+} and Pb^{2+} in its type I pocket and Zn^{2+} in its type II pocket, but only the type I binding is used for metallo-dependent derepression of the P_{cadC} promoter [Ye et al., 2005]. By contrast, the related SmtB protein from *Synechococcus* has only type II sites for Zn^{2+} binding [Turner et al., 1996].

A recent study re-engineered a number of heavy metal responsive circuits in *E. coli*, *P. fluorescens* and Grampositive bacteria like *S. aureus* or *B. subtilis* [Ivask et al., 2009]. The parts for the construction of the heavy metal responsive circuits originated from the mercury resistance operon of *S. marcescens* (*mer*, Figure 2.10), the *E. coli* chromosome (*cue/cop* for copper resistance and *znt* for zinc resistance), *C. metallidurens* CH34 (*pbr*, lead resistance) and *P. putida* PaW85 (*cad*, cadmium resistance). Circuits were produced on plasmids in *E. coli*, *P. fluorescens* and Grampositives, and in mono copy using mini-Tn5 delivery for *P. fluorescens*. All constructs used the *luxCDABE* genes of *Photobacter luminescens* as reporter, producing bioluminescence output. Interestingly, the responses of the various reporters are overlapping to a large extent, which is due to the fact that the regulatory proteins controlling the responses react similarly to groups of metal cations (See Annexes). Often, Zn^{2+}, Cd^{2+} and Pb^{2+} are not differentiated because of lack of selectivity by the regulatory protein; the same holds for Cu^+ and Ag^+, or AsIII and SbIII [Silver and le, 2005]. For further details on the various (heavy) metal responsive bioreporters, see Magrisso et al. [2008].

2.4 NETWORK INTERCEPTION DESIGNS

2.4.1 GENERAL MOTIVATION

A large number of bioreporter constructions for toxicity and mutagenicity measurements are based on stress networks in *E. coli* or *Salmonella typhimurium* [Belkin, B., 2003]. The idea behind this is that any compound or condition eliciting the stress response can be monitored with the aid of an appropriate reporter configuration. Even though such stress reporters lack compound selectivity, because many different types of compounds and conditions will lead to induction of the stress network, they benefit from being able to react very 'broadly' to anything disturbing the cell [Van Dyk et al., 2001]. Consequently, such stress reporters have raised much interest as systems to monitor the general toxicity of samples or, as explained in Chapter 1, the potential mutagenicity of compounds [Biran et al., 2009]. Some detection selectivity can be gained by interrogating different stress networks in the cell, the most important of which so far have been the SOS response, oxidative stress pathways and chaperone-induced pathways (e.g., heat-shock) [Belkin, B., 2003, Biran et al., 2009]. For an overview of different reporter constructions produced, see the Annexes. It should be noted that interrogation of global networks by transcriptional fusions can lead to small surprises because of the multiple layers of post-transcriptional control typically endowed on global networks. As a further remark, it should also be noted that despite the fact that stress network reporters are not 'compound' specific, some detection specificity can be inferred by the use of multiple parallel (different) stress reporter constructs from the global pattern of responses of the different constructs to compound exposure [Kuang et al., 2004, Lee et al., 2005, Pedahzur et al., 2004].

2.4.2 SOS RESPONSE NETWORK INTERCEPT DESIGN

The main concept in design of stress responsive bioreporters is the intercept-type (outlined above). Typical stress networks orchestrate a large number of individual genes and gene clusters scattered throughout the genome (e.g., the SOS network with at least 50 operons) [Wade et al., 2005], via the hierarchical action of one or a few central regulators (Figure 2.12). Not all of those are necessarily transcription activators, as was described for heavy metal or organic compound responsive circuits. For example, the SOS response is controlled by LexA and RecA; LexA being a transcriptional repressor binding to so-called LexA-boxes (Figure 2.13) and preventing high transcription in the absence of the SOS signals [Fry et al., 2005, Wade et al., 2005]. Conditions that lead to the formation of single-stranded DNA breaks and stalling of the replication forks will trigger RecA-mediated cleavage of LexA and loss of LexA-binding to its binding sites [Fry et al., 2005]. As a consequence, many genes are derepressed and a program is initiated of DNA repair (e.g., UvrAB), cell growth arrest (SulA), or DNA recombination (RecA, RecN). Promoters implicated in the SOS response are characterized by the presence of a consensus LexA box, but sequence and promoter architecture differences determine the magnitude of derepression at the various SOS promoters [Wade et al., 2005]. In most of the SOS reporter designs, the configuration and sequence of the chosen promoter is taken for granted (see, e.g., [Van Dyk et al., 2001]) and simply fused to a promoterless reporter gene, but in some

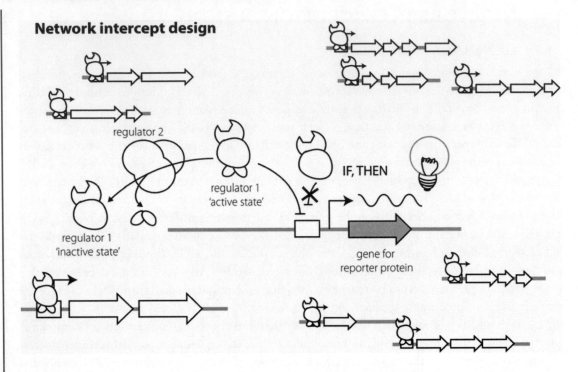

Figure 2.12: Concept of a network interception design. Imagine a large regulon, such as the SOS response, in which one regulator (e.g., LexA) controls expression of a number of different operons simultaneously. Activation of the network can be interrogated by choosing one of the network node promoters or an artificial optimized network promoter, fusing this to a promoterless reporter gene and placing the construct back into the host cell with the network. Protease cleavage of LexA (in this example) will impede its binding to the SOS promoter and elicit expression of the reporter gene.

cases, optimized constructions were produced, involving selection or engineering of more effective LexA binding sites. As an example, the *sulA* promoter has been frequently used to engineer SOS reporters [Norman et al., 2005, Yagur-Kroll et al., 2009], and it was subject to mutation analysis. Yagur-Kroll and coworkers noticed that longer promoter fragments including a portion of the *sulA* gene had a tenfold better response ratio than shorter fragments [Yagur-Kroll et al., 2009]. Whereas the length of the fragment *au fond* did not change the LexA binding site itself, other mutations were created that would do so (Figure 2.13). Interestingly, however, only mutations outside the LexA binding box further improved response ratio of this promoter (e.g, sulA18, sul-35), but not one made within (sul-10). This suggested that not only the LexA binding box itself determines an optimal SOS-reporter signal. In contrast to these results, Norman et al. compared response ratios of the *sulA*, *recA*, *umuCD*, and *cda* promoters and observed that the *cda* promoter produced the best results in

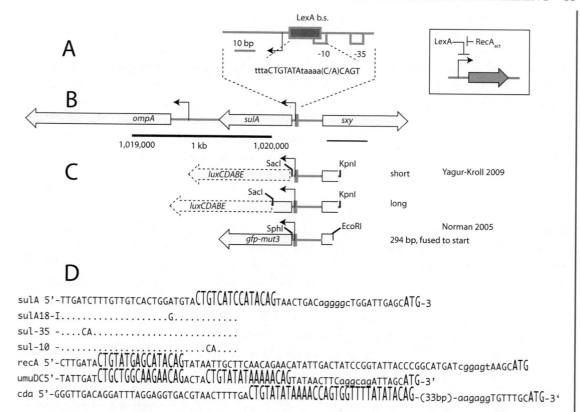

Figure 2.13: Designs and construction of LexA-controlled SOS bioreporters. Natural configuration of the *sulA* gene in *Escherichia coli* (chromosome numbering of the K12 genome) and positioning of the LexA box in the *sulA* promoter (A, B). (C) Constructions by Yagur-Kroll et al. [2009] and Norman et al. (D) Relevant sequence elements in the sulA promoters (LexA box set in capital letters), the mutant *sulA* promoters and, for comparison, the *recA*, *umuDC* and *cda* promoters [Norman et al., 2005].

gene reporter assays [Norman et al., 2005]. They attributed this to a double-long extended LexA binding box (Figure 2.13).

2.5   PROMOTER ENGINEERING

2.5.1   GENERAL NOTIONS

As outlined above, the first choice for a synthetic sensor/reporter construction often starts with selecting the sensory-regulatory protein with the required target specificity and a suitable control region for tuning expression of the reporter signal. From what I explained above, we can also conclude that in most sensor/reporter constructions so far, researchers have applied the native configuration of the control region 'as is', without further specific efforts to make rational designs. Furthermore, because of the ease of molecular biology toolbox in *E. coli*, many designs have been assembled on plasmid DNA in this species. If we restrain ourselves to the four sets of examples described in the preceding Sections (e.g., ArsR, MerR, HbpR and XylR), we can also observe that in most cases the genes for the sensory-regulatory protein were included on the same genetic construction in the cell and being under control of their own native promoters. In three of those designs, the gene for the sensor-regulator protein is divergently oriented from the reporter gene. The only exception to this is *arsR*, which is co-oriented with the reporter gene. The obvious reason for this was to remain as close as possible to the native system, although there has been no systematic investigation as to whether this makes a design better or worse. In the case of the *arsR* system, it has become clear that the co-orientation leads to significant background expression of the reporter gene if no measures are taken downstream of *arsR* to prevent read-through [Stocker et al., 2003]. Sensor/reporter constructions based on MerR produce such a low limit of Hg^{2+} detection (see Annexes) that there is no reason (yet) to re-engineer the close positions of the various elements in this control region. On the other hand, XylR-based designs have typically assembled the *xylR* gene on the same construction as the P_U control region, which do not naturally occur together, and therefore have implicitly made engineering decisions to combine control elements together (Figure 2.7). Incidentally, in producing a new genetic delivery system for constructions in *Pseudomonas*, de las Heras and coworkers 2008 noted that the 47-bp FRT recombination site for the yeast flippase enzyme also acted as a sort of transcriptional shield for background expression coming from P_{S1} and P_U in the absence of any chemical effector (Figure 2.7). Despite still being scarce, a number of specific attempts have been made to rationally optimize or change the control element in the sensor/reporter design, which I will discuss in more detail.

2.5.2   PROMOTER ENGINEERING IN THE ARS SYSTEM

Let us first turn again our attention on the ArsR-based designs, which suffer from relatively high background expression of the reporter gene when this is directly coupled downstream of *arsR* (Figure 2.11). Because there is no transcription terminator downstream of *arsR*, placing promoterless reporter genes downstream of *arsR* will result in their expression being controlled by the ArsR regulated promoter upstream of *arsR*. For bioassays with bacterial strains carrying such sensor/reporter constructs read-through is not automatically a disadvantage when the signal-to-noise ratio remains sufficiently high even at arsenic concentrations close to the method of detection limit, but this feature of the *arsR* system in practice prevents sensitive detection at low As concentrations. In our own research, we thus decided to find an approach to control read-through of transcription through *arsR* into the reporter gene but still maintain arsenite-inducible expression. Placing a terminator downstream of *arsR* would not solve the problem because this would also dislodge arsenite-inducible expression of the reporter gene. Neither would an uncoupling of *arsR* and reporter gene expression have immediately solved the problem, for example, by orienting *arsR* and its control region divergently from the reporter gene. In this case, one would again have to use an ArsR-regulated promoter in front of the reporter gene, for which currently the only one known is P_{ars} itself. We thus entertained the hypothesis that we might block transcriptional read-through by using ArsR itself. If ArsR would also attach on the DNA downstream of *arsR* and in front of the reporter gene, it might abort RNA polymerase to continue transcription into the reporter gene. This would not *per se* influence expression of *arsR* itself, which is necessary for the circuit to work. We thus decided to place an extra copy of the ArsR binding site, but not the -10 and -35 regions of P_{ars} downstream of *arsR* (Figure 2.11). Indeed, this effectively lowered background expression of the reporter gene to a level making colorimetric and fluorimetric tests more sensitive. In contrast, despite lowering background expression, quantitative measurements using bacterial luciferase as reporter were not showing a better dynamic range [Stocker et al., 2003].

In an attempt to further tune arsenite-inducible expression of the reporter gene in the config-uration of having a secondary ArsR binding site downstream of *arsR*, we then engineered different control elements involving mainly the character of the ribosome binding site (RBS) for *lacZ* (Fig-ure 2.11). By removing the native *lacZ* RBS, but recreating a poor RBS (GGAAG) between -4 and -9 upstream of the start codon, we achieved a sensor/reporter construct, which had a rela-tively high method of detection limit for arsenite in corresponding bioassays (≈ 50 μg As[III] per L) [Wackwitz et al., 2008]. In contrast, by including the complete native *lacZ* upstream region but without P_{LAC}, an extremely sensitive sensor/reporter construct was obtained. Bioassays with *E. coli* cells containing this construct resulted in a method of detection limit of less than 0.1 μg As per L [Wackwitz et al., 2008]. This showed that by engineering additional control elements one can further optimize an existing 'native' regulatory circuit, and manipulating the RBS sequence or its position is certainly one good strategy in this.

2.5.3   PHYSIOLOGICAL CONTROL OF THE XYLR-REGULATED P_U PROMOTER

The next example is a summary of engineering designs to remove physiological control on the XylR-regulated P_U promoter. In this strategy, the group around Victor de Lorenzo decided to construct hybrid promoters between P_U and a number of other σ^{54}-dependent promoters such as *nifHp* and *glnPA2p* [Carmona et al., 2005]. The design thus consisted of maintaining the sequence and position of the proximal and distal UAS-boxes for XylR in P_U, but replacing the 160-bp fragment containing the IHF binding site and the actual -12, -24 sequences of P_U by the corresponding ones from the *nifHp* and *glnPA2p* (Figure 2.7). In this case, the gene for *xylR* was not included in the same genetic circuit but was present elsewhere in the genome of the sensor/reporter strain. As a matter of fact, hybrid promoters kept inducibility by XylR and toluene, but, in particular, the P_U /*glnPAp* hybrid was devoid of the exponential phase silencing effect [Carmona et al., 2005]. This engineering was made before the discovery of the aforementioned PprA binding site that overlaps with the XylR UASs in the P_U promoter [Vitale et al., 2008]. Therefore, it is not completely clear which control elements were effectively removed from P_U that exert exponential phase silencing control. Already earlier on, the team around de Lorenzo demonstrated how spacing effects between the two UASs and the σ^{54}-promoter, or DNA sequences producing intrinsically curved structures, can influence the overall strength of the P_U promoter 1994. This, however, to my knowledge, was not further applied in sensor/reporter concepts.

2.5.4   DUAL RESPONSIVE CONTROL SWITCHES

In my own research group, we decided to investigate whether it would be possible to obtain dual responsive promoters by combining control elements for two different sensor-regulators. Hereto we chose the HbpR and XylR systems because their binding sites were relatively well understood (Figure 2.8). Since the native UASs of HbpR and XylR are relatively similar, but not identical, we constructed a number of different UAS pairs within the P_C control region, which gradually changed either UAS-C2 for HbpR into UAS-D (U2) for XylR, or UAS-C1 into UAS-P (or U1) [Tropel et al., 2004]. In doing so, we noticed that a hybrid couple of UAS containing UAS-C2 and UAS-U1 was activated both by HbpR and XylR. Interestingly, also control regions carrying both UAS-P and UAS-D sequences but within an otherwise P_C sequence context, were inducible to higher level than the native P_U, perhaps because they lack the PprA binding site (Figure 2.8). Not only was the sensor/reporter construct pHB244 containing the hybrid P_C with UAS-C2 and UAS-P activated by HbpR in the presence of 2-hydroxybiphenyl or by XylR in the presence of m-xylene in strains with either $hbpR$ or $xylR$, but also in a strain containing both $xylR$ and $hbpR$ [Tropel et al., 2004]. This showed that dual control can be achieved and also showed that re-engineering regulator binding sites in a different sequence background can remove additional control factors, such as exponential phase silencing.

2.5.5   DIRECTED EVOLUTION OF PROMOTERS

Although this is currently more important for metabolic engineering, several groups have approached the question as to how obtain promoters of different 'strength' to drive gene expression but without the need of induction by a chemical effector [Alper et al., 2005, De Mey et al., 2007]. In contrast to existing theory that dictates that the positions and sequences of -10 and -35, plus anomalies with respect to the optimal spacer length between them (17 nt) would be decisive for promoter strength, newest results suggest that this is not the case. De Mey et al. [2007] recently developed a model based on Partial Least Square analysis of nucleotide positions to predict promoter output. Although this exercise has not been directly applied for sensing/-reporting, where the primary purpose is to quantify a target compound, the engineering principles for promoter strength are important to understand. The use of controllable constitutive promoters with different strength might also be conceivable for driving expression of the sensor-regulatory protein itself in order to avoid auto-feedback loops (as for $arsR$ and P_{ARS}), or of auxiliary factors (e.g., compound influx or efflux transporters). Experimentally, promoter evolution was approached via random mutagenesis or random promoter synthesis of short fragments based on constitutive promoters such as P_L from phage lambda, coupled to, e.g., gfp expression or cat selection (for chloramphenicol acetyl transferase activity) [Alper et al., 2005]. Evolved promoters with a higher activity would be detectable because they result in brighter cells or colonies, and those with lower activity in less brightly fluorescent cells or colonies. The major outcome of this type of work was that it is indeed possible to obtain a range of promoter sequences that are seemingly independent of *cis* and *trans*-acting factors, and result in different target gene expression [Alper et al., 2005, De Mey et al., 2007]. By using gfp as reporter, differential expression of up to 1000-fold was observed, which translated into promoters having 30 times more or 7 times less output than P_{LAC} or P_L [Alper et al., 2005, De Mey et al., 2007].

A different approach to studying cis-acting elements important for promoter output consisted of designed combinatorial libraries of promoter elements (originating from various described binding sites for four activators and repressors) fused to the *yfp* (yellow fluorescent protein) reporter gene [Kinkhabwala and Guet, 2008]. To control the combinatorial promoters a combination of four transcriptional regulators was chosen, three of which were constitutively expressed (LacI and TetR repressors and the AraC activator), and one of which was under inducible expression (λcI under P_{BAD} and arabinose). Interestingly, the authors found that the reporter output logic from the combinatorial promoters was more or less independent from binding site strength of individually chosen elements [Kinkhabwala and Guet, 2008]. This path of integrating different inputs into useful reporter output has not been described for sensor/reporter constructions, as far as I am currently aware, but it is now much studied in the context of expression noise and transcriptional logic.

2.5.6   RESPONSE HETEROGENEITY IN POPULATIONS

Inherent to biological systems is the heterogeneity of responses among individuals in a population. This is no different for biosensor/-reporter strains, the result of which is that not all cells respond with the same reporter signal intensity despite the fact that, in theory, they should all see the same chemical effector concentration (we will see in Chapter 3 that this assumption may not always be true). Depending on the detection system of the reporter signal, such heterogeneity of response could become a nuisance. Imagine, for example, a population of bacterial sensor/reporter cells with a Normal distributed reporter protein activity (Figure 2.14). If one were to measure the integrated output of the whole population, the reporter signal would correspond to the mean, whereas if one would analyze the reporter signal in every individual cell (e.g., by epifluorescence microscopy), one would find that it is Normal distributed around the mean. More heterogeneous expression in the population of cells would lead to the same mean but individual cells would display considerable more reporter signal variability. This does not necessarily pose any problems for measurements, except when analyzing very small numbers of reporter cells. A major problem, though, can arise with so-called Bimodal populations, in which individual cells display one of two response states of the reporter. This behavior occurs for wild-type systems like *lac* and *ara* (for lactose and arabinose metabolism, respectively). At suboptimal inducer concentrations, some cells in a population are not at all induced whereas others are already induced to a maximum level. This was confusingly described as an 'all or none' phenomenon, which in essence has nothing to do with the regulatory control region, but with the fact that the transporter system for the inducer is under the same control in the

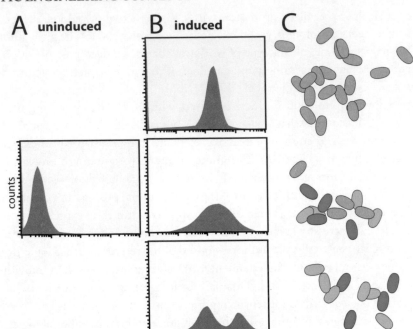

Figure 2.14: The reporter response in all cells of a population can be heterogeneous to a different degree. In the top example, a rather homogenous response of all individuals is taking place resulting in a small standard deviation of the reporter signal distribution around the mean. In the middle example, the response of the cells is more heterogeneous, but the mean and integrated output of the whole population remains the same as in the top panel example. In the lower panel, a Bimodal response is shown in which some cells hardly and others very strongly react. The mean and integrated output of the whole population can still remain the same as in the other two exampleBimodal but the underlying biological behavior of the lower example is completely different.

cognate system [Khlebnikov et al., 2000, Morgan-Kiss et al., 2002]. The result of this is a feedback amplification loop and strongly heterogeneous response of the cells to the inducer. Interestingly, for the widely used arabinose inducible system (e.g., pBAD vectors), heterogeneity of expression in individual cells can be overcome by uncoupling arabinose transport in an *araE araFGH* strain (*araE* overexpression) [Khlebnikov et al., 2000] or by using a relaxed LacY transporter in an *ara* background [Morgan-Kiss et al., 2002]. Noisyness of promoters has a clear biological function and, for example, a simple screen was developed to identify promoters, which are more heterologously expressed in cells than others [Freed et al., 2008]. For other systems, the heterogeneity is determined by the internal and external noise (see Section 2.6).

It is important to note that the expression of the reporter protein is not the same thing as transcription from a promoter, and that whereas most researchers in the field of sensor/reporter engineering use reporter output to measure promoter strength or promoter response variability, a variety of posttranscriptional mechanisms can influence optimal protein synthesis. Miller and Lindow noticed poor *gfp* expression from initial promoter probe vectors that was due to suboptimal placement of the ribosome binding site and probably unfavorable 5' secondary folding of the transcript [Miller and Lindow, 1997]. They engineered an optimized translation initiation region (TIR) in front of the *gfp* gene, which improved expression from weaker promoters by fourfold [Miller and Lindow, 1997]. The engineered spacer region was a combination of a region with poor secondary structure folding, optimal spacing of the ribosome binding site to the start codon and an enhancer sequence originating from the gene 10 of phage T7. The issue of posttranscriptional behavior was further recently illustrated in a synthetic attempt to understand which factors would determine optimal gfp expression in *E. coli* [Kudla et al., 2009]. The authors expressed codon-randomized gfp's in *E. coli*, but found that highly expressed gfp's did not display any particular codon usage pattern. Instead, it appeared that low folding energy of the most 5' 50 nucleotides of the mRNA (-4 to +37 relative to the start codon) explained most of the variation in optimal gfp expression. This suggested that poorly folding sequences in this part of the messenger would facilitate ribosomal access and protein translation. To illustrate the importance of their finding further, a synthetic low folding tag sequence was produced (introducing 28 codons at the 5' end) that could be fused directly downstream of the start codon and upstream of the normal *gfp* sequence. This tag dramatically improved the homogeneity and the level of gfp expression in a population of *E. coli* cells from typical bacterial promoters such as P_{LAC} and P_{BAD} [Kudla et al., 2009].

2.6 ENGINEERING NEW EFFECTOR SPECIFICITIES

2.6.1 GENERAL CONCEPT

As I outlined above, of primary essence to the design of a biosensor/-reporter system is the choice of the protein that has to sense the chemical target. In the examples presented earlier, the choice for the sensing protein was made implicitly; in taking a bacterium degrading toluene, one expected to have a sensing protein for toluene (e.g., XylR); in using the mercury defense system, one expected to find a sensing protein for mercury (e.g., MerR), and so forth. Obviously, this expectation was not always correct: in the case of naphthalene or 2,4-dichlorophenoxyacetic acid metabolism, it is not the target compound but an intermediate that is the actual chemical *effector* for the regulatory circuit (e.g., salicylate in naphthalene degradation and NahR; *cis, cis*-muconate in the case of 2,4-dichlorophenoxyacetic acid and TfdR). In addition, it was noticed that in most cases the spectrum of effectors for a particular sensing protein encompassed more than one specific chemical compound. For example, XylR can react to a whole suite of compounds (toluene, *m*- and *p*-xylene, 3-methylbenzylalcohol, 3-aminotoluene 2006), and the ArsR protein not only reacts to arsenite (As[III]), but also to antimonite (Sb[III]) and even wolframite (W[III]) albeit with much less sensitivity [Wu and Rosen, 1993]. Would it, therefore, not be an excellent idea if one could tailor the effector specificities of a sensory/regulatory protein to, firstly, narrow the amount of chemical effectors, and, secondly, to compounds which normally do not act as effectors? A few groups in bacterial reporter research embarked on this prospect and indeed attained new effector specificities for their model proteins. On the other hand, conceptually it appeared to be quite challenging to conclude whether these changes really involved new specificities or an expansion from a pre-existing one [Galvao and de Lorenzo, 2006, Galvao et al., 2007, Garmendia et al., 2001]. Ideally, one would like to be able to rationally project the evolutionary trajectory of such sensory-regulatory proteins from one effector recognition state to the other.

2.6.2   EFFECTOR DOMAIN MUTAGENESIS IN XYLR/DMPR-TYPE PROTEINS

The examples which I would like to present here to illustrate the concepts and types of effector changes on sensory-regulatory proteins include the XylR, DmpR and HbpR proteins. They are also particularly interesting because this group of proteins all fall within the same class and demonstrate a relatively clear subdomain architecture, one of which forms the supposed sensing domain (i.e., the A-domain) [Morett and Segovia, 1993, Shingler, V., 1996]. The role of the A-domain was postulated on the basis of spontaneous and induced mutations in this part of the protein, which abolished effector-inducible transcription activation [Delgado et al., 1995, Fernández and de Lorenzo, 1995, Fernández et al., 1994, Shingler and Moore, 1994]. In addition, direct binding of radioactively labeled effector compound (^{14}C-phenol) to the A-domain part of the protein was shown for DmpR [O'Neill et al., 1998, 1999]. Final piece of evidence for the hypothesized effector recognition was that complete removal of the A-domain for XylR and DmpR resulted in proteins that did no longer respond to their chemical inducers, but retained constitutive transcription activation [Fernández and de Lorenzo, 1995, Ng et al., 1996]. Despite these results however, there is still no precise notion of what the effector-binding pocket in this class of proteins is constituted of, although A-domain models have been proposed on the basis of weak homologies to structure-resolved proteins [Devos et al., 2002].

The strategy to construct and find mutants with altered effector recognition specificities was then the following: isolate the gene fragment for the A-domain part of the protein, produce a library of mutant A-domain gene fragments by directed evolution strategies, replace the pool to reconstitute a complete regulator gene library and select (or screen) for effector induction. This approach was followed more or less in similar fashion for DmpR, XylR and HbpR, whereby the directed evolution consisted of an error-prone type DNA amplification of the gene fragment for the A-domain or DNA shuffling between two different A-domain gene fragments (for example, one from XylR, the other from DmpR) [Beggah et al., 2008, Galvao and de Lorenzo, 2006]. Where the approaches differed, however, was in the manner of selecting or screening for the potential effector mutants. A screening approach was demonstrated for DmpR A-domain mutations and consisted of coupling the activity of the regulator protein to drive expression of a reporter gene, such as beta-galactosidase or luciferase, from its cognate promoter (called P_O). The library of potential mutants was then plated on media containing different types of potential chemical effectors [Skarfstad et al., 2000, Wikstrom et al., 2001, Wise and Kuske, 2000], and colonies with brighter reporter signals were screened for further analysis. This strategy worked and although the number of colonies that could be screened was modest ($<10^4$), DmpR variants with altered effector specificities were recovered [Sarand et al., 2001, Wise and Kuske, 2000]. Interestingly, however, most of such mutants actually retained wild-type recognition but expanded their capability to become induced by other effectors.

A similar strategy but with higher throughput screening was followed by our own group, in which we attempted to obtain effector mutants of the aforementioned protein HbpR that would recognize chlorinated biphenyls instead of hydroxylated ones [Jaspers et al., 2000]. The reason for attempting this was that there is currently no bacterium that seems to possess a regulator protein to directly sense chlorinated biphenyls, whereas such compounds are considered to be important environmental pollutants for which monitoring is essential. A mutant library of HbpR A-domains was thus created by error-prone PCR, reconstituted to the full-length gene and used in *E. coli* to drive expression of *gfp* from the cognate HbpR-dependent P$_C$ promoter [Beggah et al., 2008]. The advantage of using gfp as reporter protein was that we could use flow cytometry to screen larger numbers of mutants, and, in the case of detection of a gfp-induced cell on 2-chlorobiphenyl, could separate it out by fluorescence assisted cell sorting (FACS) to culture it. Also this approach was to some extent successful, although it was underestimated (as in the case of DmpR) how often mutations lead to a constitutive phenotype. Constitutive in this sense means cells that express gfp to the same level, irrespective of the presence or absence of the chemical effector tested, or cells that displayed a very high gfp level in the absence of effector, which increased slightly more after exposure of the mutant to the effector. Such phenomena led the team of de Lorenzo to propose that evolutionary trajectories of effector recognition might go through stages in which the regulatory protein reacts (semi-) constitutively, before new recognition specificity is fully gained and constitutive behavior is again diminished [Galvao et al., 2007]. As before for DmpR, the most interesting mutants obtained for HbpR with responsiveness to 2-chlorobiphenyl still responded to 2-hydroxybiphenyl as well, although at least one was isolated, which was less strongly activated by 2-hydroxy- than by 2-chlorobiphenyl [Beggah et al., 2008].

To overcome the potential difficulties in screening large numbers of mutants by classical reporter gene phenotypes, a different approach was taken in which action of the regulatory protein was coupled to a killing of the cells during specific culture conditions [Galvao, 2005]. One of the developed systems consisted hereto of a gene for orotidine-5'-phosphate decarboxylase (*pyrF*), which is essential for producing uracil. Strains lacking *pyrF* cannot grow on media without uracil, but they can be complemented by placing *pyrF* under control of the XylR-dependent P$_U$-promoter. Media containing effectors for XylR would then lead to induction of PyrF and cells can grow [Mohn et al., 2006]. Interestingly, PyrF-expression can also kill cells when they are plated on media containing the uracil analogue 5-fluoro-orotic acid. The strategy was applied to select for XylR mutants with increased responsiveness to 2,4-dinitrotoluene (a byproduct from explosives). An A-domain library of some 10^4 mutants was screened on media with 5-fluoro-orotic acid to eliminate constitutive XylR mutants because these would not grow on such conditions. Survivors were then screened for growth on plates without uracil but with 2,4-dinitrotoluene [Garmendia et al., 2008].

2.6.3   EFFECTOR BINDING POCKET MODELING

The gold standard, obviously, would be to have such a detailed insight on the structure of the regulatory protein and the effector binding pocket that one could use molecular modeling to predict the necessary residue changes to arrive at a new effector recognition. Such an approach was tested for the DntR protein, which regulates expression of the *dnt* pathways for dinitrotoluene degradation in *P. putida*. The crystal structure of DntR is available and, therefore, this research group was interested to change the effector recognition of DntR from the native effector salicylate to 2,4-dinitrotoluene by molecular modeling predictions and site-directed mutagenesis [Smirnova et al., 2004]. A little disappointingly, the approach was not (yet) very successful, and engineered mutants were little better than the wild-type DntR in responding to 2,4-dinitrotoluene [Lonneborg et al., 2007]. It appeared that although the new binding pocket was predicted to fit 2,4-dinitrotoluene, it was 'distorted' by a slightly different overall structure that was adopted by the mutant protein [Lonneborg et al., 2007]. It is to be expected, though, that approaches based on molecular modeling will become more powerful and successful in the near future when protein folding can be predicted more accurately.

2.6.4   APTAMERS

A completely different strategy for obtaining effector specificity in a reporter circuit is to use aptamers or riboswitches, typically RNAs with strong and specific three-dimensional structures accepting small molecules or changing conformation upon binding a small molecules. Aptamers can be produced by *in vitro* directed evolution techniques such as SELEX (systematic evolution of ligands by exponential enrichment) [Klug and Famulok, 1994], generating large variant libraries from which molecules with the appropriate substrate binding properties can be selected. However, whereas apparent substrate binding affinity constants for aptamers and small molecules can easily be in the low μM range, this does not automatically make them good biological elements for gene reporter circuits. In order to embed an aptamer in a gene reporter circuit, one would need an activation or derepression mechanism to unleash reporter expression. Activating aptamers would need to prevent reporter gene expression (often by posttranscriptional mechanisms) in the absence of effector, but would undergo a conformational change upon binding an effector molecule thereby promoting expression. Alternatively, in the unbound state, they would allow reporter expression but prevent it after binding the effector. For a recent example in the light of gene reporter circuitry, Sinha et al. engineered aptamer libraries for binding the herbicide atrazine [Sinha et al., 2010]. They selected not only atrazine-binding but 'switchable' aptamers by coupling the aptamer sequence to the *cheZ* chemotaxis regulator gene, and screened for swarming cells on soft agar plates with atrazine. A number of atrazine-'inducible' aptamer sequences were retrieved, and induction could be demonstrated by coupling the aptamer to a LacZ reporter. Induction was caused by more effective translation of the reporter mRNA when atrazine was bound to the aptamer sequence, as opposed to induction of transcription as in most of the reporter circuitry examples presented here. Although apparent affinity constants for *in vitro* atrazine binding to the aptamer sequence were in the order of 2 μM, cells with the engineered aptamer-reporter gene only responded in the range of 100 to 500 μM atrazine with up to fivefold increased reporter signal. This makes the atrazine aptamer a relatively insensitive bioreporter but so far an interesting example of how new detection specificities may be engineered and perhaps be improved in the future [Sinha et al., 2010].

2.7   COMPLEX SIGNAL-TRANSDUCTION CHAINS

2.7.1   GENERAL CONCEPT

Most examples outlined above used relatively simply sensor/reporter constructions in which a single transcription regulator protein is responsible for sensing of the target chemical and for eliciting transcription from the cognate promoter. Implicit in the use of such sensor/reporter constructions is that the target chemical will diffuse or be transported into the cytoplasm in order to be detected by the transcription regulator. Very few designs have attempted to take advantage of more complex signal transduction systems that most bacteria possess. A number of designs have used the two component regulatory system TodST from *P. putida*, which is responsible for control of expression of a toluene degradation pathway [Applegate et al., 1998, Lovanh and Alvarez, 2004, Shingleton et al., 1998]. Toluene and a wide range of other aromatic effectors are sensed by the N-terminal part of the TodS protein. Upon effector binding, autophosphorylation of TodS is enhanced and the phosphorylated form of TodS subsequently transphosphorylates the response regulator protein TodT. Phosphorylated TodT binds to the *todX* promoter, but for optimal activation of RNA polymerase, it further needs Integration Host Factor (IHF) to configure a specific bend on the promoter DNA [Lacal et al., 2006]. The TodST system has orthologues in a number of other bacterial genomes, of which so far only the related StyRS system has been exploited for sensor/reporter constructions ([Alonso et al., 2003], see Annex).

2.7.2   PERIPLASMIC BINDING PROTEINS AND PHOSPHOTRANSFER RELAY

In an attempt to possibly intercept target chemicals not in the cytoplasm but in the periplasm, the group around Hellinga pioneered with designs using periplasmic binding proteins (PBPs) in *E. coli* [Looger et al., 2003]. PBPs are frequently found proteins that bacterial cells use to scavenge certain sugars or amino acids in the periplasmic space and present the captured molecules to specific transporters. In a few cases, such as ribose-binding protein (RBP) or galactose-binding protein (GBP) in *E. coli*, the PBP-captured sugar is docked onto a membrane chemoreceptor (for RBP and GBP this is the Trg receptor) that can direct the chemotactic response of the cell [Hazelbauer et al., 2008]. Unfortunately, the chemotactic response of a cell consists of a phosphotransfer relay that does not involve *de novo* gene expression [Stock et al., 2000] and is therefore not something directly exploitable for sensor/reporter constructions. However, already a while ago it was demonstrated that hybrid membrane histidine kinases can be constructed that have the chemosensory part of the chemoreceptor (like Trg) fused to the membrane spanning and cytoplasmic domains of a classical two-component histidine kinase such as EnvZ [Baumgartner et al., 1994, Stock et al., 2000]. The hybrid between Trg and EnvZ (called Trz) was thus capable of sensing sugar bound RBP or GBP and transducing the phosphorylation via the EnvZ-core to the cognate response regulator of EnvZ, which is OmpR [Stock et al., 2000]. Phosphorylated OmpR, on its turn, is capable of activating the *ompC* promoter, which can be fused to a reporter gene and thus serves as signal output for sugar perception [Baumgartner et al., 1994]. Whereas this concept of hybrid sensing was thus proposed a while ago, the innovation of the design came with the realization that PBPs (of which crystal structures are known) can be remodeled to accommodate non-natural substrates. This was demonstrated in the work by Looger et al., who showed that several different PBPs can be computationally designed, then produced by site-directed mutagenesis to capture non-trivial molecules like serotonin or L-lactate [Looger et al., 2003]. More specifically for RBP and GBP, they demonstrated the computational design of variants detecting trinitrotoluene, which were synthesized and embedded in an *E. coli* bioreporter strain producing the hybrid Trz-OmpR-P_{ompC}-*lacZ* response [Looger et al., 2003]. Further computational modeling of such docking PBPs may thus prove a very fruitful effort in the future to overcome the problem of finding transcription factors with the appropriate chemical target binding properties in the classical gene reporter designs (Figure 2.15). More recently, an iGEM group at Brown university used the same concept to computationally propose a structure of a variant RBP binding histamine and couple this via the Trz receptor cascade to production of rfp (red fluorescent protein) in *E. coli*[5].

2.8   MULTINODE NETWORKS

2.8.1   LOGIC GATES, TRANSCRIPTIONAL NOISE, AMPLIFICATION

Much recent interest concerns the construction of multinode reporter networks, building logic gates on the basis of regulatory networks, and the understanding of noise in gene networks. Most of this work is fundamentally oriented but with the long-term purpose of building expression networks in a similar manner as, e.g., building electrical circuits. Unfortunately, most research groups active in this area have used a limited number of very standardized transcription regulators, such as LacI, TetR, AraC and λcI (see below), which are not very exciting for sensing purposes. The earlier cited work by Kinkhabwala explored different regulator and promoter binding site configurations to build logic gates [Kinkhabwala and Guet, 2008]. Figure 2.16 illustrates a number of designs that have been produced.

In the so-called 'repressilator', Elowitz and Leibler produced a number of feedback loops in series, which lead to oscillating formation of an (unstable) gfp reporter over time in cells in a culture [Elowitz and Leibler, 2000]. The essential part in this design was not only the construction of the feedback loops, but also the fusion of tags to the various transcription factors and

[5]http://2009.igem.org/Team:Brown/Project_Histamine_Sensor

Membrane receptor design

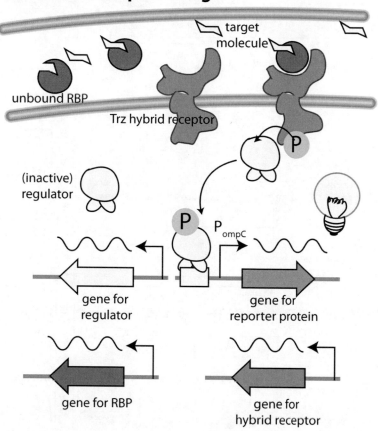

Figure 2.15: Sensor/reporter design exploiting periplasmic binding proteins and a hybrid sensory kinase. Ribose binding protein (RBP) or an engineered variant with modified substrate binding properties intercepts the target in the periplasmic space. In bound form the RBP docks to the sensor domain of the Trg receptor and triggers autophosphorylation of the hybrid EnvZ cytoplasmic part. This leads to phosphorylation of OmpR and OmpR-P activates the *ompC* promoter, which is fused to the reporter gene. Concept after [Baumgartner et al., 1994, Looger et al., 2003].

Cascades

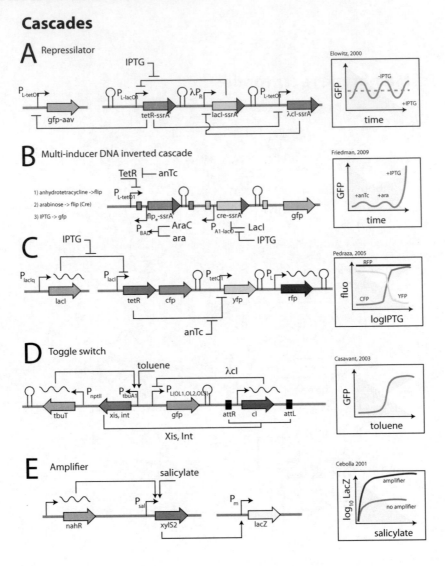

Figure 2.16: Examples of complex reporter layouts. (A) Repressilator, leading to oscillatory behavior of the reporter gene in single cells [Elowitz and Leibler, 2000]. (B) Cell counters that produce reporter output only after a predefined number of times of contact to (a set of) inducers [Friedland et al., 2009]. (C) Network to study noise propagation in a defined range of inducer concentration [Pedraza, P., 2005]. (D) Toggle-switch, leading to maximum reporter output after signal perception; useful when measuring the number of cells with reporter output rather than reporter intensity over the culture [Casavant et al., 2003]. (E) Signal amplification circuit, involving expression of a secondary constitutively active regulator [Cebolla et al., 2001].

reporter protein that would increase protein degradation and allow observation of the oscillation (Figure 2.16 (A)). The system is solely controlled by the addition of IPTG as effector for LacI-SsrA derepression on the $P_{L-lacO1}$ promoter. Oscillation of gfp fluorescence could be observed in single (synchronized) microcolonies of cells, whereas oscillation was disrupted by the addition of 50 μM IPTG [Elowitz and Leibler, 2000].

In the Pedraza design of 2005 [Pedraza, P., 2005], a number of loops were produced in series in order to study the propagation of noise in a small network (Figure 2.16 (C)). In this design LacI is constitutively expressed and controlling IPTG-dependent expression of *tetR* and a *cfp* reporter placed in series. TetR, on its turn, is repressing a *yfp* reporter gene, repression of which can be relieved by the addition of anhydrotetracycline (anTc). Finally, an independent constitutively placed *rfp* gene serves to monitor constant output of the system. At low IPTG concentrations LacI will effectively repress *tetR* and *cfp* transcription, leading to YFP formation but not CFP (Figure 2.16 (C)). At high concentrations LacI will no longer repress *tetR* transcription and TetR, on its turn, will inhibit *yfp* expression. In this case, cells will be brightly cfp but low yfp. In the intermediate situation of 10 to 100 μM IPTG cells will start to behave noisy and some of them will produce yfp but not cfp, whereas others will show cfp and not yfp, depending on the state of the network in an individual cell [Pedraza, P., 2005].

Friedland developed the concept of a cell counter, in which bacterial cells will display different output depending on whether they have been exposed once, twice or thrice to pulses of the inducer arabinose, or to a sequential addition of anhydrotetracycline, arabinose and IPTG [Friedland et al., 2009]. In the layout displayed in Figure 2.16, a gene for an instable yeast flippase is repressed by TetR, but addition of anTc leads to derepression from this promoter and synthesis of the flippase, which will lead to inversion of the DNA with the flippase gene in between the target recombination sites. Flipping will uncouple the flippase gene from its promoter but place the AraC controlled P_{BAD} promoter in front of a gene coding for the Cre recombinase. Addition of arabinose will now activate expression of Cre, which will turn the DNA with its own gene around and place a LacI controlled promoter in front of the *gfp* reporter gene. Addition of IPTG will derepress the $P_{A1-lacO}$ promoter and lead to formation of gfp. Hence, the cells have counted to three before forming larger amounts of the gfp reporter protein. Also in this design, essential additions included protein degradation tags to the flippase and Cre to counteract infrequent secondary 'flipping' effects that would bring the respective genes again back under control of the original promoters. The authors described a second concept using T7 and T3 RNA polymerase, the expression of which was controlled by *cis*-acting mRNA repressor sequences that can be targeted by a *trans* activator RNA. Expression of the *trans* activator RNA was brought under control of the arabinose-dependent P_{BAD} promoter [Friedland et al., 2009].

The idea of a recombination-dependent cascade had been exploited before by Casavant et al. to build a toluene 'all or nothing' reporter [Casavant et al., 2003]. The concept behind this construction was not so much to have a reporter in which the reporter signal would be relevant for the target chemical concentration, but the amount of cells in a population displaying the (maximum) signal.

This reporter, so the authors, would be useful in more complex environments like soil or plant roots [Casavant et al., 2003]. In their design expression of the Xis/Int recombinase is under control of the toluene-dependent TbuT transcription activator. Expression of Xis/Int will cut out a piece of DNA containing the gene for the Lambda cI repressor that in absence of toluene will repress transcription of *gfp* from a P_L promoter (Figure 2.16 (D)). More or less independent of the toluene concentration, exposed cells will remove λcI and maximally express gfp.

A final useful concept to achieve signal amplification was explored by Cebolla et al. [2001]. Bacterial bioreporters for environmental chemicals often display relatively low reporter output when inducer concentrations are limiting (see, e.g., [Tecon et al., 2009]), and one of the methods to overcome this would be to amplify the generated primary signal. The authors demonstrated this concept by coupling the primary signal (salicylate-dependent NahR activation) to expression of a secondary regulatory gene (*xylS2*) instead of a reporter gene. XylS2 is capable of constitutively activating expression from the P_M-promoter and the authors could show that salicylate exposure leads to much higher LacZ reporter gene expression in cells carrying the amplification circuit compared to a single circuit in which *lacZ* is directly fused to the P_{SAL} promoter (Figure 2.16 (E)).

2.9 REPORTER PROTEINS

2.9.1 CHOICE AND SPECIFICITIES OF COMMON REPORTER PROTEINS

An essential part of a gene reporter circuitry is the reporter protein itself. Much has been written about various reporter proteins, their specificities and usage, and it is not my primary goal to repeat all this [Biran et al., 2009, Daunert et al., 2000, Hakkila et al., 2002, Kohler et al., 2000, Leveau and Lindow, 2002, Magrisso et al., 2008, Shaner et al., 2004, 2008]. On the other hand, no text on bioreporter engineering can go without a description of various reporter proteins, and a summary of some of the most used reporters with their main characteristics is presented in Table 2.1. The good thing about reporter proteins is that one is practically 'spoiled'; fluorescent proteins and luciferases exist in various colors, making it possible to combine various reporters in one cell, tailor output to appropriate devices, and so on. The downside of this perhaps is that the choice of suitable reporter becomes more and more difficult, and appropriate quantification in reporter assays not automatically more easily. One can see from the information in the Annexes on various produced bioreporter constructs that most groups have used luciferase (eukaryotic or prokaryotic), for reasons of sensitivity and selectivity.

The essence of a reporter protein is that it allows more sensitive, accurate and easy measurement of the 'reaction' of a cell than the cognate output. Ideally, a reporter protein is not naturally present in the host cell for the reporter construction because this may dramatically increase the background activity/presence of the reporter signal one wishes to distinguish. Also, measurement of the reporter protein's activity or presence should ideally be done in a 'window' where no confounding other cellular signals are present. For example, fluorescence of a reporter protein may be disturbed by autofluorescence of other cellular components in the same excitation-emission wavelengths. Then, the reporter protein should obviously not be toxic for the producer cell or else its presence disturbs the reaction one wishes to measure and interpret. Native dsRed, for example, has been relatively difficult to express in bacteria without considerable impact on growth rates, something that has been surmounted with the engineering of a new generation of red fluorescent proteins (e.g., mCherry [Shaner et al., 2004]). Finally, a reporter protein ideally allows non-invasive measurement of its activity or presence to minimize disturbance to the study system. As a consequence, of all this, many proteins have been proposed as reporter and probably not a single one is ideal for all purposes. Initial reporter proteins were chosen because they permitted color reactions after addition of substrates, some of which required cell permeabilization to obtain the desired color reaction (e.g., LacZ beta-galactosidase and ONPG, see Table 2.1). Other reporter proteins permitted easy phenotypic screening tests even in complex samples but without the possibility to observe at single-cell level (e.g., Ice-nucleation protein). The currently most widely used reporter proteins involve fluorescence or bioluminescence, which allow highly sensitive and precise measurements, but sometimes with the drawback of demanding specific conditions to become active (e.g., oxygen).

The choice for an appropriate reporter protein is therefore a balance between availability of instrumentation, signal sensitivity, need for additional substrates, ease of cloning, reporter interference or stability in the cell, the possibility for performing multiple consecutive reporter measurements or direct interference of environmental conditions on the functioning of the reporter. In some cases, people may want to use combinations of different reporter proteins in the same cell, for example, to measure two different signals simultaneously or to have one constitutively expressed reporter and one inducible. A procedure that has been followed in several toxicity bioreporters is to have a constitutive luciferase signal and a stress-inducible fluorescent reporter. Since the activity of the luciferase is sensitive to compounds that disturb the energy state of the cell, one has an internal control to determine the presence of any toxic substance that may hinder the proper development of the inducible reporter signal. For these and other examples, see Biran et al. [2009]. In our own work, we have had very good experience with combinations such as egfp plus mCherry, which display very different excitation and emission wavelengths and, therefore, do not present much risk of over-lapping signals and false interpretations [Tecon et al., 2009]. We have also performed comparative quantitative measurements with the same genetic circuit but different reporter genes (e.g., bacterial luciferase, egfp, beta-galactosidase and yeast perodixase) [Kohlmeier et al., 2007, Stocker et al., 2003, Wackwitz et al., 2008] and with the same reporter gene and different analysis methods [Wells et al., 2005]. This showed that whereas, in general, luciferase and beta-galactosidase allow lower method of detection limits, the precision of the reporter signal measurement mostly depends on the type of detection instrument and experimental layout [Wells et al., 2005].

2.9.2   REPORTER GENE VECTORS

Many groups will have experienced cloning frustrations in producing the appropriate gene reporter construction, once the choice has been made for a suitable reporter gene. One reason for this is that a multitude of personalized vectors are around, which in specific cases exactly do not allow the cloning one is intending to produce. Table 2.1 and Figure 2.17 present some information on reporter vectors ('promoter probe vectors') that various groups including our own have been using or proposing, although this listing is by no means extensive. As one can see, some vectors allow exclusive expression in *E. coli*, whereas others permit engineering in *E. coli* and subsequent transfer to an alternative host strain of interest. Such vectors typically have two origins of replication and a mobilization region, or exploit a chromosomal delivery system using, e.g., the mini-Tn5 [Herrero, H., 1990, Kristensen et al., 1995] or mini-Tn7 transposases [Lambertsen et al., 2004]. Certain vectors allow some sort of flow chart cloning in the sense of producing the construct in one vector, then recovering the whole assembly of promoter and reporter genes as one unique fragment using the seldom cutting enzymes NotI (SfiI or PacI), and subsequently inserting those fragments in a more complicated (i.e., allowing limited manipulation) second vector (e.g., Figure 2.17 (G)), from where the reporter circuit can be inserted into the chromosome. Most vectors are designed for transcriptional fusions to a promoter region of interest, and have, for example, included stop codons in all frames and specific ribosome binding sites or translation enhancers directly upstream of the reporter gene (e.g., pPROBE series, Table 2.1 [Miller et al., 2000]). However, as the discussion above on translational elements implied, one should be aware of possible posttranscriptional regulatory effects on reporter circuits, or implicate such specifically. There are current efforts to further streamline various cloning vectors for a wide range of bacterial hosts and re-engineer mini-transposon vectors for easier cloning (Victor de Lorenzo, personal information). Direct DNA synthesis will likely more and more replace classical cloning and engineering efforts, in which case precise design of the construction into a vector of choice is largely facilitated. The most comprehensive repository of sequence building blocks with reporter genes, promoters, terminators, or further transcriptional and translational elements is currently http://parts.mit.edu/registry, although this one is getting so full that finding the building block one is looking for is far from being simple.

2.9.3   EMBEDDING OF SENSING/-REPORTING CIRCUITS IN A CELLULAR CHASSIS

A final word in this Chapter concerns the host cell or 'cellular chassis' for the reporter circuit. As discussed in the beginning of the Chapter, there may be various different reasons for creating reporter circuits. As I proposed, one could differentiate between fully orthogonal constructions (calle/operating 'outside' the original cognate cellular reaction) and the interception design, in which a cellular reaction or network is interrogated as closely as possible to the natural behavior. Interception designs, therefore, may have to be constructed in the original host cell of which one likes to determine the reaction to a signal. For example, the aforementioned SOS response of *E. coli* is best measured in *E. coli* itself. Producing an orthogonal version in another host would likely interfere with a similar SOS response in that host and result in the 'loss' of the original interpretation of the *E. coli* behavior. In another example, one may like to study bioavailability of phenanthrene, which happens not to be an easy trait to reproduce in *E. coli*. Since phenanthrene metabolism is required for measuring phenanthrene bioavailability with the help of a reporter construct, one would in first instance be 'forced' to turn to the original host and implant the reporter there (with all possible difficulties of

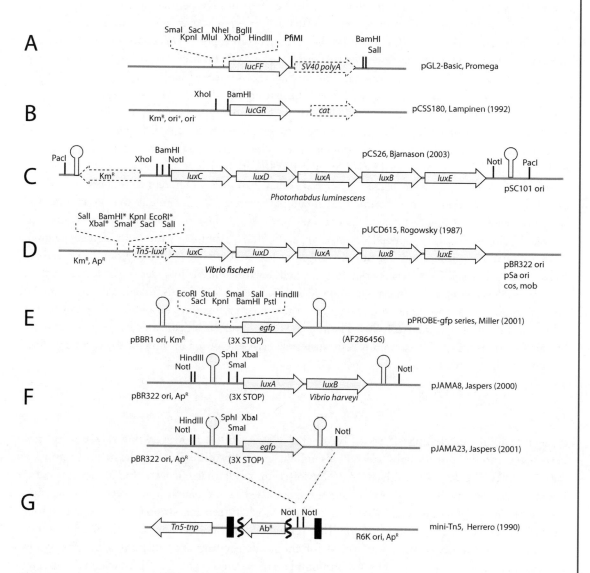

Figure 2.17: Caption appears on the next page.

Figure 2.17: Schematic outline of a number of reporter vectors that have been frequently used in reporter circuit engineering. (A) Basic firefly luciferase reporter vector from Promega for use in *E. coli* and with an extensive choice of restriction sites. Further variants exist permitting optimal expression of eukaryotic promoters and in eukaryotic hosts. (B) Luciferase reporter vector for use in Grampositive hosts [Lampinen et al., 1992]. (C) Vector for bacterial luciferase reporter (*Photorhabdus luminescens*) and internal substrate generation developed by Bjarnason et al. [2003] for use in *E. coli* and *Salmonella*. Note inclusion of transcription terminators and unique flanking sites (NotI and PacI) for subsequent flow chart cloning. (D) Luciferase reporter vector with the bacterial luciferase and substrate generation genes from *Vibrio fischeri* [Rogowsky et al., 1987]. The vector has two origins of replication, for *E. coli* and for *Agrobacterium*, into which it can be mobilized by conjugation. Restriction sites marked with an asterisk are unique in the vector. Note that the region upstream of *luxC* carries a fragment of a Tn5 inserted into luxI. (E) pPROBE series developed by Miller et al. [2000] with both *egfp* or *inaZ* as reporter. Vectors are broad-host range and can be mobilized into species other than *E. coli*. Different resistances are available, as well as different orientations of the multiple cloning sites. For egfp, different unstable variants are available. The sequence of this vector is available from Genbank (AF286456). (F) pJAMA series of reporter vectors developed in our own group [Jaspers et al., 2000, 2001b]. Simple vectors with high copy number in *E. coli* that allow transcriptional fusions to *egfp* or bacterial luciferase (*luxAB* from *Vibrio harveyi*), transcriptionally shielded by two terminators up- and downstream. Vectors were designed as such that subsequent flow chart cloning into a mini-Tn5 system via unique NotI sites is feasible. (G) The mini-Tn5 systems allow monocopy chromosomal delivery of the reporter construct in a number of host bacteria other than *E. coli* [Herrero, H., 1990]. The region within black bars delineates the region transposed by the Tn5 transposase. The antibiotic resistance gene can be removed in some of the vectors via ParA resolution (zig-zag lines) [Kristensen et al., 1995].

non-amenability to cloning efforts!) [Tecon et al., 2009, 2006]. However, one sees that in many cases even interception designs are not assembled in exactly the wild-type host. For example, it was early noted that the sensitivity of *E. coli* of compounds that elicit the SOS response can be boosted by knocking out the genes for UvrA-mediated excision repair and for polysaccharide capsid synthesis (*raf*) [Quillardet et al., 1982]. In addition, toxicity-sensing strains often carry mutations in *tolC* to cause an inactivation of the TolC efflux system [Biran et al., 2009]. Such mutations increase the sensitivity of the tester strains for chemicals eliciting toxic responses, but this ignores the original self-defense of the strain to compound toxicity [Biran et al., 2009]. To widen the apparent spectrum of compounds that would elicit a toxicity response, research groups are more and more including metabolic genes into the *E. coli* or *Salmonella* tester strains, for example, those expressing mammalian P450 oxygenases [Aryal et al., 2000]. This would circumvent the use of rat liver extract exogenously added to a reporter assay to transform certain chemicals to more mutagenic forms (see Chapter 1) [Biran et al., 2009].

Hence, one sees how sensitivity of the reporter strain to the chemical target becomes one of the most important issues to achieve, but it should be clear that this leads away from an interception idea (the natural response of an organism) to an orthogonal design ('no matter what, the most sensitive possible reaction'). Host strain engineering has also played a role in bioreporters reactive to heavy metals, mostly for two reasons: (i) to improve sensitivity of the reporter cell to low concentrations of heavy metals, and (ii) to try and create a more selective response. We have seen above in Section 2.3 how metal reporters create specificity via the sensory/regulator protein, but one of the important cellular mechanisms in conjunction with the regulatory protein-inducible response to heavy metals are transporters. Metals, because of their charge, typically have to be transported inside the cell where the reporter circuit is operating. In addition, they are often rapidly removed by active efflux systems or by a chemically modified, precipitated or complexed by variety of other enzymatic mechanisms [Silver and le, 2005]. As an example of the effect of the host strain on the performance of different metal reporter circuits, see Ivask et al. [2009]. They not only recorded mildly different detection limits and sensitivities for various metals in four host strains *E. coli*, *P. fluorescens*, *Staphylococcus aureus* and *Bacillus subtilis*, but found that in terms of toxicity, the strains react quite differently [Ivask et al., 2009].

A different purpose for modifying host strains as recipients in reporter constructions is that of the synthetic engineering itself. As mentioned earlier, going outside the regular laboratory strains of *E. coli* can be very significant in terms of environmental or toxic response, but very unrewarding in terms of ease of assembling the reporter circuit in all its details. One example for this, which is particularly relevant for sensing/reporter purposes at present, is performed by the pioneering work of groups such as around Victor de Lorenzo and around Herbert Schweizer, who have engineered endless sets of vectors and systems in especially pseudomonads (but functioning in related bacteria). Recently, de las Heras and coworkers proposed a refined form of a universal reporter docking site integrated in monocopy on the chromosome of the host *Pseudomonas putida* [de las Heras et al., 2008]. The advantage of the docking site would be that it only has to be engineered once, upon which different regulatory circuits can be placed directly upstream of it using mini-Tn7 transposition (Figure 2.18). In addition, the system is equipped as such that antibiotic resistance genes used for construction and selection purposes can be removed from the final strain, thereby minimizing undesired gene fragments in the final host. An additional advantage of the designed docking system is that it permits to use the Tn7 attachment site in host species where this sequence is not naturally present, whereas the Tn5 transposase that is used to insert the docking site is more promiscuously active in a wide variety of hosts [de las Heras et al., 2008].

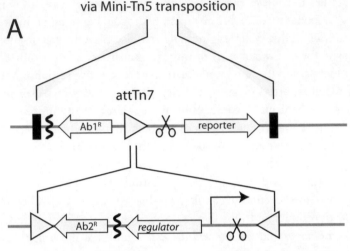

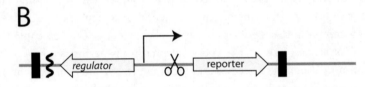

Figure 2.18: Principle of creating a single copy universal reporter docking site into which different activation or repression circuits can be inserted [de las Heras et al., 2008]. At the first instance, the docking site is created in *E. coli*, which containts the reporter gene, a Tn7 attachment site (triangle), an antibiotic resistance gene, one copy of the ParA resolvase site (the zig-zag line), and one copy of the yeast flippase FRT recombination site (scissor); all of which is flanked by Tn5 transposase recognition sites (black bars). The docking site is inserted in mono-copy into the genome of interest by mini-Tn5 transposition. After selection of the proper insertion the regulatory circuit is assembled and integrated at the Tn7 attachment site using mini-Tn7 transposition. After selection of the proper Tn7 insertion into the docking site, the antibiotic resistance genes are removed step-by-step through site-specific resolution (ParA) and recombination (yeast flippase), leaving only the regulatory circuit in place in front of the reporter gene. The only 'scars' are one copy of the yeast flippase gene, one copy of the ParA resolvase site and the mini-Tn5 boundaries.

Table 2.1: Reporter proteins. Key: BL, bioluminescence; FL, fluorescence; COL, colorimetry; EC, electrochemistry; CL, chemiluminescence; RI, radioactivity; HIS, histology; IN, ice nucleation. Asterisks point to where internal substrate for the enzyme is generated. *Continues.*

Reporter	Output	Substrate (s)	Advantages	Disadvantages	Plasmid source	Reference
LuxAB (V.h.) Bacterial luciferase	BL	n-decanal, O_2, (FMNH$_2$)*	Highly sensitive	Needs substrate addition. Needs specific instrument	pJAMA8	(Jaspers, Suske et al. 2000)
LuxCDABE (V.f.) Bacterial luciferase (biosynthesis)	BL	O_2, (none)*	Highly sensitive. No substrate addition	Needs specific instrument. Internal substrate generation prone to external disturbance. Large gene fragment to clone. Not very stable at 37°C.	pUCD615	(Rogowsky, Close et al. 1987)
LuxCDABE (P.l.) Bacterial luciferase (biosynthesis)	BL	O_2, (none)*	As for LuxCDABE Vf, but more stable at 37°C	As for LuxCDABE Vf	pTET-Lux1 pCS26	(Frackman, Anhalt et al. 1990) (Bjarnason, Southward et al. 2003)
Gfp (A.v.) Green fluorescent protein	FL	O_2	No substrate addition. Non-invasive, single cell level. Various instrument possibilities	Not very sensitive, slow maturation, extremely stable. No signal amplification as for enzymatic reporter protein.		

Table 2.1: *Continued.* Reporter proteins. Key: BL, bioluminescence; FL, fluorescence; COL, colorimetry; EC, electrochemistry; CL, chemiluminescence; RI, radioactivity; HIS, histology; IN, ice nucleation. Asterisks point to where internal substrate for the enzyme is generated. *Continues.*

Reporter	Output	Substrate (s)	Advantages	Disadvantages	Plasmid source	Reference
Egfp Engineered autofluorescent protein	FL	O_2	As for gfp. Improved maturation and photostability. Wide variety of colors exist	Needs oxygen for maturation. No signal amplification as for enzymatic reporter protein	pPROBE-tagless pJAMA23	(Miller, Leveau et al. 2000) (Jaspers, Meier et al. 2001)
mCherry Engineered autofluorescent protein	FL	O_2	As for egfp, multiple colors exist	Needs oxygen for maturation	pJAMA39	(Minoia, Gaillard et al. 2008)
LucFF Firefly luciferase	BL	O_2, ATP, luciferin	Extremely sensitive, no endogenous activity	Needs substrate addition, (for bacteria:) permeabilization of cells	pGL2-Basic	http://www.prome ga.com.cn/techser v/tbs/TM00Ⱶ 310/tm003.pdf
lucGR Click beetle luciferase	BL	O_2, ATP, phosalin	As lucFF, different color variants	As lucFF		(Cebolla, Vazquez et al. 1995)
InaZ Ice nucleation protein	IN		No substrates needed, no specific instruments needed	Only endpoint measurement	pPROBE-GI/KI/NI	(Miller, Leveau et al. 2000)
XylE Catechol 2,3-dioxygenase	COL	O_2, catechol	Cheap substrate	Poor color, endogenous activity	pNM185	(Mermod, Ramos et al. 1986)

Table 2.1: *Continued.* Reporter proteins. Key: BL, bioluminescence; FL, fluorescence; COL, colorimetry; EC, electrochemistry; CL, chemiluminescence; RI, radioactivity; HIS, histology; IN, ice nucleation. Asterisks point to where internal substrate for the enzyme is generated.

Reporter	Output	Substrate (s)	Advantages	Disadvantages	Plasmid source	Reference
LacZ Beta-galactosidase	COL, HIS, FL, CL, EC	Galactopyranoside substrate	Extremely versatile, highly sensitive, does not require oxygen	Endogenous activity, requires substrate addition, sometimes requires cell permeabilization (ONPG)	pPZ10, pPZ20, pPZ30, X1918, Z1918	(Schweizer 1991; Schweizer 1993)
GusA. β-D-glucuronidase	COL, FL, HIS, CL	Glucorinide substrate	Wide variety of substrates. Most often used as reporter in higher plants.	Endogenous activity in bacteria.	pNZ272	(Platteeuw, Simons et al. 1994)
CrtA Spheroidene monooxygenase	COL	Spheroidene*	No equipment needed for detection of color change	Endogenous background, less widely used	pSENSE-AS	(Fujimoto et al., 2006)
Yeast cytochrome oxidase	COL, CL	Guiacol	Various mutants exist	Color reaction not brilliant	pET15b-CCP	(Wachwitz et al., 2008)

REFERENCES

Alonso, S., D. Bartolome-Martín, et al. (2003). "Genetic characterization of the styrene lower catabolic pathway of *Pseudomonas* sp. strain Y2." *Gene* 319: 71–83. DOI: 10.1016/S0378-1119(03)00794-7 43

Alper, H., C. Fischer, et al. (2005). "Tuning genetic control through promoter engineering." *Proc. Natl. Acad. Sci. U. S. A.* 102(36): 12678–12683. DOI: 10.1073/pnas.0504604102 36, 37

An, W. and J. W. Chin (2009). "Synthesis of orthogonal transcription-translation networks." *Proc. Natl. Acad. Sci. U. S. A.* 106(21): 8477–8482. DOI: 10.1073/pnas.0900267106 12

Applegate, B. M., S. R. Kehrmeyer, et al. (1998). "A chromosomally based *tod-luxCDABE* whole-cell reporter for benzene, toluene, ethybenzene, and xylene (BTEX) sensing." *Appl. Environ. Microbiol.* 64(7): 2730–2735. 43

Arenghi, F. L., M. Pinti, et al. (1999). "Identification of the *Pseudomonas stutzeri* OX1 toluene-o-xylene monooxygenase regulatory gene (*touR*) and of its cognate promoter." *Appl. Environ. Microbiol.* 65(9): 4057–4063. 19

Aryal, P., T. Terashita, et al. (2000). "Use of genetically engineered *Salmonella typhimurium* OY1002/1A2 strain coexpressing human cytochrome P450 1A2 and NADPH-cytochrome P450 reductase and bacterial O -acetyltransferase in SOS/ umu assay." *Environ. Mol. Mutagen.* 36: 121.126. DOI: 10.1002/1098-2280(2000)36:2%3C121::AID-EM6%3E3.0.CO;2-P 52

Baumgartner, J. W., C. Kim, et al. (1994). "Transmembrane signalling by a hybrid protein: communication from the domain of chemoreceptor Trg that recognizes sugar-binding proteins to the kinase/phosphatase domain of osmosensor EnvZ." *J. Bacteriol.* 176(4): 1157–1163. 44, 45

Beggah, S., C. Vogne, et al. (2008). "Mutant transcription activator isolation via green fluorescent protein based flow cytometry and cell sorting." *Microb. Biotechnol.* 1: 68–78. 24, 40, 41

Belkin, S. (2003). "Microbial whole-cell sensing systems of environmental pollutants." *Curr. Opin. Microbiol.* 6(3): 206–212. DOI: 10.1016/S1369-5274(03)00059-6 31

Biran, A., R. Pedahzur, et al. (2009). "Genetically engineered bacteria for genotoxicity assessment." *Biosensors for the Environmental Monitoring of Aquatic Systems*. D. Barcelo and P.-D. Hansen. Berlin/ Heidelberg, Springer: 161–186. DOI: 10.1007/978-3-540-36253-1_6 31, 48, 49, 52

Bjarnason, J., C. M. Southward, et al. (2003). "Genomic profiling of iron-responsive genes in Salmonella enterica serovar typhimurium by high-throughput screening of a random promoter library." *J. Bacteriol.* 185(16): 4973–4982. DOI: 10.1128/JB.185.16.4973-4982.2003 52

Carmona, M., S. Fernandez, et al. (2005). "m-xylene-responsive Pu-PnifH hybrid σ^{54}-promoters that overcome physiological control in *Pseudomonas putida* KT2442." *J. Bacteriol.* 187(1): 125–134. DOI: 10.1128/JB.187.1.125-134.2005 22, 35

Casavant, N. C., D. Thompson, et al. (2003). "Use of a site-specific recombination-based biosensor for detecting bioavailable toluene and related compounds on roots." *Environ. Microbiol.* 5: 238–249. DOI: 10.1046/j.1462-2920.2003.00420.x 46, 47, 48

Cebolla, A., C. Sousa, et al. (2001). "Rational design of a bacterial transcriptional cascade for amplifying gene expression capacity." *Nucl. Acids. Res.* 29: 759–766. DOI: 10.1093/nar/29.3.759 46, 48

Cebolla, A., M. E. Vazquez, et al. (1995). "Expression vectors for the use of eukaryotic luciferases as bacterial markers with different colors of luminescence." *Appl. Environ. Microbiol.* 61(2): 660–668.

Chen, Y. and B. P. Rosen (1997). "Metalloregulatory properties of the ArsD repressor." *J. Biol. Chem.* 272(22): 14257–14262. DOI: 10.1074/jbc.272.22.14257 28

Collard, J. M., P. Corbisier, et al. (1994). "Plasmids for heavy metal resistance in *Alcaligenes eutrophus* CH34: mechanisms and applications." *FEMS Microbiol. Rev.* 14(4): 405–414. DOI: 10.1111/j.1574-6976.1994.tb00115.x 30

Corbisier, P., G. Ji, et al. (1993). "*luxAB* Gene Fusions with the Arsenic and Cadmium Resistance Operons of *Staphylococcus aureus* Plasmid-pI258." *FEMS Microbiol. Lett.* 110(2): 231–238. DOI: 10.1111/j.1574-6968.1993.tb06325.x 30

Daunert, S., G. Barrett, et al. (2000). "Genetically engineered whole-cell sensing systems: coupling biological recognition with reporter genes." *Chem. Rev.* 100(7): 2705–2738. DOI: 10.1021/cr990115p 13, 48

de las Heras, A., C. A. Carreno, et al. (2008). "Stable implantation of orthogonal sensor circuits in Gram-negative bacteria for environmental release." *Environ. Microbiol.* 10(12): 3305–3316. DOI: 10.1111/j.1462-2920.2008.01722.x 20, 22, 34, 53, 54

de Lorenzo, V., I. Cases, et al. (1993). "Early and late responses of TOL promoters to pathway inducers: identification of postexponential promoters in *Pseudomonas putida* with *lacZ-tet* bicistronic reporters." *J. Bacteriol.* 175: 6902–6907. 21

de Lorenzo, V., M. Herrero, et al. (1991). "An upstream XylR- and IHF-induced nucleoprotein complex regulates the σ^{54}-dependent Pu promoter of TOL plasmid." *EMBO J.* 10(5): 1159–1169. 19

De Mey, M., J. Maertens, et al. (2007). "Construction and model-based analysis of a promoter library for *E. coli*: an indispensable tool for metabolic engineering." *BMC Biotechnol.* 7: 34. DOI: 10.1186/1472-6750-7-34 36, 37

Delgado, A., R. Salto, et al. (1995). "Single amino acids changes in the signal receptor domain of XylR resulted in mutants that stimulate transcription in the absence of effectors." *J. Biol. Chem.* 270(10): 5144–5150. DOI: 10.1074/jbc.270.10.5144 40

Devos, D., J. Garmendia, et al. (2002). "Deciphering the action of aromatic effectors on the prokaryotic enhancer-binding protein XylR: a structural model of its N-terminal domain." *Environ. Microbiol.* 4(1): 29–41. DOI: 10.1046/j.1462-2920.2002.00265.x 40

Diaz, E. (2004). "Bacterial degradation of aromatic pollutants: a paradigm of metabolic versatility." *Int. Microbiol.* 7(3): 173–180. 17

Diaz, E. and M. A. Prieto (2000). "Bacterial promoters triggering biodegradation of aromatic pollutants." *Curr. Opin. Biotechnol.* 11(5): 467–475. DOI: 10.1016/S0958-1669(00)00126-9 14

Elowitz, M. B. and S. Leibler (2000). "A synthetic oscillatory network of transcriptional regulators." *Nature* 403(6767): 335–338. DOI: 10.1038/35002125 44, 46, 47

Fernández, S., V. de Lorenzo, et al. (1995). "Activation of the transcriptional regulator XylR of *Pseudomonas putida* by release of repression between functional domains." *Mol. Microbiol.* 16: 205–213. DOI: 10.1111/j.1365-2958.1995.tb02293.x 40

Fernández, S., V. Shingler, et al. (1994). "Cross-regulation by XylR and DmpR activators of *Pseudomonas putida* suggests that transcriptional control of biodegradative operons evolves independently of catabolic genes." *J. Bacteriol.* 176(16): 5052–5058. 40

Frackman, S., M. Anhalt, et al. (1990). "Cloning, organization, and expression of the bioluminescence genes of *Xenorhabdus luminescens*." *J. Bacteriol.* 172(10): 5767–5773.

Freed, N. E., O. K. Silander, et al. (2008). "A simple screen to identify promoters conferring high levels of phenotypic noise." *PLoS Genet.* 4(12): e1000307. DOI: 10.1371/journal.pgen.1000307 38

Friedland, A. E., T. K. Lu, et al. (2009). "Synthetic gene networks that count." *Science* 324(5931): 1199–1202. DOI: 10.1126/science.1172005 46, 47

Fry, R. C., T. J. Begley, et al. (2005). "Genome-wide responses to DNA-damaging agents." *Annu. Rev. Microbiol.* 59: 357–377. DOI: 10.1146/annurev.micro.59.031805.133658 31

Fujimoto, H., M. Wakabayashi, et al. (2006). "Whole-cell arsenite biosensor using photosynthetic bacterium *Rhodovulum sulfidophilum*. *Rhodovulum sulfidophilum* as an arsenite biosensor." *Appl. Microbiol. Biotechnol.* 73(2): 332–338. DOI: 10.1007/s00253-006-0483-6

Galvao, T. C. and V. de Lorenzo (2005). "Adaptation of the yeast URA3 selection system to gram-negative bacteria and generation of a δbetCDE *Pseudomonas putida* strain." *Appl. Environ. Microbiol.* 71(2): 883–892. DOI: 10.1128/AEM.71.2.883-892.2005 41

Galvao, T. C. and V. de Lorenzo (2006). "Transcriptional regulators a la carte: engineering new effector specificities in bacterial regulatory proteins." *Curr. Opin. Biotechnol.* 17(1): 34–42. DOI: 10.1016/j.copbio.2005.12.002 17, 39, 40

Galvao, T. C., M. Mencia, et al. (2007). "Emergence of novel functions in transcriptional regulators by regression to stem protein types." *Mol. Microbiol.* 65(4): 907–919. DOI: 10.1111/j.1365-2958.2007.05832.x 40, 41

Garmendia, J., A. de las Heras, et al. (2008). "Tracing explosives in soil with transcriptional regulators of *Pseudomonas putida* evolved for responding to nitrotoluenes." *Microb. Biotechnol.* 1: 236–246. DOI: 10.1111/j.1751-7915.2008.00027.x 42

Garmendia, J. and V. de Lorenzo (2000). "Visualization of DNA-protein intermediates during activation of the Pu promoter of the TOL plasmid of *Pseudomonas putida*." *Microbiology* 146(Pt 10): 2555–2563. 19

Garmendia, J., D. Devos, et al. (2001). "A la carte transcriptional regulators: unlocking responses of the prokaryotic enhancer-binding protein XylR to non-natural effectors." *Mol. Microbiol.* 42(1): 47–59. DOI: 10.1046/j.1365-2958.2001.02633.x 40

Hakkila, K., M. Maksimow, et al. (2002). "Reporter Genes *lucFF, luxCDABE, gfp*, and *dsred* Have Different Characteristics in Whole-Cell Bacterial Sensors." *Anal. Biochem.* 301: 235–242. DOI: 10.1006/abio.2001.5517 48

Hay, A. G., J. F. Rice, et al. (2000). "A bioluminescent whole-cell reporter for detection of 2,4-dichlorophenoxyacetic acid and 2,4-dichlorophenol in soil." *Appl. Environ. Microbiol.* 66: 4589–4594. DOI: 10.1128/AEM.66.10.4589-4594.2000 18

Hazelbauer, G. L., J. J. Falke, et al. (2008). "Bacterial chemoreceptors: high-performance signaling in networked arrays." *Trends. Biochem. Sci.* 33(1): 9–19. DOI: 10.1016/j.tibs.2007.09.014 43

Herrero, M., V. de Lorenzo, et al. (1990). "Transposon vectors containing non-antibiotic resistance selection markers for cloning and stable chromosomal insertion of foreign genes in Gram-negative bacteria." *J. Bacteriol.* 172(11): 6557–6567. 50, 52

Ivask, A., K. Hakkila, et al. (2001). "Detection of organomercurials with sensor bacteria." *Anal. Chem.* 73(21): 5168–5171. DOI: 10.1021/ac010550v 26

Ivask, A., T. Rolova, et al. (2009). "A suite of recombinant luminescent bacterial strains for the quantification of bioavailable heavy metals and toxicity testing." *BMC Biotechnol.* 9: 41. DOI: 10.1186/1472-6750-9-41 30, 53

Jaspers, M. C., A. Schmid, et al. (2001a). "Transcriptional organization and dynamic expression of the *hbpCAD* genes, which encode the first three enzymes for 2-hydroxybiphenyl degradation in *Pseudomonas azelaica* HBP1." *J. Bacteriol.* 183(1): 270–279. DOI: 10.1128/JB.183-1.270-279.2001 22

Jaspers, M. C., W. A. Suske, et al. (2000). "HbpR, a new member of the XylR/DmpR subclass within the NtrC family of bacterial transcriptional activators, regulates expression of 2-hydroxybiphenyl metabolism in *Pseudomonas azelaica* HBP1." *J. Bacteriol.* 182(2): 405–417. DOI: 10.1128/JB.182.2.405-417.2000 19, 22, 24, 41, 52

Jaspers, M. C. M., C. Meier, et al. (2001b). "Measuring mass transfer processes of octane with the help of an *alkS-alkB::gfp*-tagged *Escherichia coli*." *Environ. Microbiol.* 3(8): 512–524. DOI: 10.1046/j.1462-2920.2001.00218.x 52

Jaspers, M. C. M., M. H. Sturme, et al. (2001c). "Unusual location of two nearby pairs of upstream activating sequences for HbpR, the main regulatory protein for the 2-hydroxybiphenyl degradation pathway of 'Pseudomonas azelaica' HBP1." *Microbiology* 147: 2183–2194. 22

Khlebnikov, A., O. Risa, et al. (2000). "Regulatable arabinose-inducible gene expression system with consistent control in all cells of a culture." *J. Bacteriol.* 182(24): 7029–7034. DOI: 10.1128/JB.182.24.7029-7034.2000 38

Kim, M. N., H. H. Park, et al. (2005). "Construction and comparison of *Escherichia coli* whole-cell biosensors capable of detecting aromatic compounds." *J. Microbiol. Meth.* 60: 235–245. DOI: 10.1016/j.mimet.2004.09.018 21

King, J. M. H., P. M. DiGrazia, et al. (1990). "Rapid, sensitive bioluminescent reporter technology for naphthalene exposure and biodegradation." *Science* 249(4970): 778–781. DOI: 10.1126/science.249.4970.778 17

Kinkhabwala, A. and C. C. Guet (2008). "Uncovering cis regulatory codes using synthetic promoter shuffling." *PLoS One* 3(4): e2030. DOI: 10.1371/journal.pone.0002030 37, 44

Klug, S. J. and M. Famulok (1994). "All you wanted to know about SELEX." *Mol. Biol. Rep.* 20(2): 97–107. DOI: 10.1007/BF00996358 42

Kohler, S., S. Belkin, et al. (2000). "Reporter gene bioassays in environmental analysis." *Fresenius J. Anal. Chem.* 366(6–7): 769–779. DOI: 10.1007/s002160051571 48

Kohlmeier, S., M. Mancuso, et al. (2007). "Bioreporters: *gfp* versus *lux* revisited and single-cell response." *Biosens. Bioelectron.* 22(8): 1578–1585. DOI: 10.1016/j.bios.2006.07.005 49

Kristensen, C. S., L. Eberl, et al. (1995). "Site-specific deletions of chromosomally located DNA segments with the multimer resolution system of broad-host-range plasmid RP4." *J. Bacteriol.* 177(1): 52–58. 50, 52

Kuang, Y., I. Biran, et al. (2004). "Living bacterial cell array for genotoxin monitoring." *Anal. Chem.* 76(10): 2902–2909. DOI: 10.1021/ac0354589 31

Kudla, G., A. W. Murray, et al. (2009). "Coding-sequence determinants of gene expression in *Escherichia coli*." *Science* 324(5924): 255–258. DOI: 10.1126/science.1170160 39

Lacal, J., A. Busch, et al. (2006). "The TodS-TodT two-component regulatory system recognizes a wide range of effectors and works with DNA-bending proteins." *Proc. Natl. Acad. Sci. U. S. A.* 103(21): 8191–8196. DOI: 10.1073/pnas.0602902103 43

Lambertsen, L., C. Sternberg, et al. (2004). "Mini-Tn7 transposons for site-specific tagging of bacteria with fluorescent proteins." *Environ. Microbiol.* 6(7): 726–732. DOI: 10.1111/j.1462-2920.2004.00605.x 50

Lampinen, J., L. Koivisto, et al. (1992). "Expression of luciferase genes from different origins in *Bacillus subtilis*." *Mol. Gen. Genet.* 232: 498–504. DOI: 10.1007/BF00266255 52

Laurie, A. D. and G. Lloyd-Jones (1999). "The *phn* genes of *Burkholderia* sp. strain RP007 constitute a divergent gene cluster for polycyclic aromatic hydrocarbon catabolism." *J. Bacteriol.* 181(2): 531–540. 19

Leahy, J. G., G. R. Johnson, et al. (1997). "Cross-regulation of toluene monooxygenases by the transcriptional activators TbmR and TbuT." *Appl. Environ. Microbiol.* 63(9): 3736–3739. 19

Lee, J. H., R. J. Mitchell, et al. (2005). "A cell array biosensor for environmental toxicity analysis." *Biosens. Bioelectr.* 21(3): 500–507. DOI: 10.1016/j.bios.2004.12.015 31

Leveau, J. H. J. and S. E. Lindow (2002). "Bioreporters in microbial ecology." *Curr. Opin. Microbiol.* 5: 259–265. DOI: 10.1016/S1369-5274(02)00321-1 9, 48

Li, Y. F., F. Y. Li, et al. (2008). "Construction and comparison of fluorescence and bioluminescence bacterial biosensors for the detection of bioavailable toluene and related compounds." *Environ. Pollut.* 152(1): 123–129. DOI: 10.1016/j.envpol.2007.05.002 30

Lloyd, G., P. Landini, et al. (2001). "Activation and repression of transcription initiation in bacteria." *Essays Biochem.* 37: 17–31. 24

Lonneborg, R., I. Smirnova, et al. (2007). "In vivo and in vitro investigation of transcriptional regulation by DntR." *J. Mol. Biol.* 372(3): 571–582. DOI: 10.1016/j.jmb.2007.06.076 42

Looger, L. L., M. A. Dwyer, et al. (2003). "Computational design of receptor and sensor proteins with novel functions." *Nature* 423: 185–189. DOI: 10.1038/nature01556 43, 44, 45

Lovanh, N. and P. J. Alvarez (2004). "Effect of ethanol, acetate, and phenol on toluene degradation activity and tod-lux expression in *Pseudomonas putida* TOD102: evaluation of the metabolic flux dilution model." *Biotechnol. Bioeng.* 86(7): 801–808. DOI: 10.1002/bit.20090 43

Magrisso, S., Y. Erel, et al. (2008). "Microbial reporters of metal bioavailability." *Microb. Biotechnol.* doi:10.1111/j.1751-7915.2008.00022.x. DOI: 10.1111/j.1751-7915.2008.00022.x 30, 48

Marques, S., M. T. Gallegos, et al. (1998). "Activation and Repression of Transcription at the Double Tandem Divergent Promoters for the *xylR* and *xylS* Genes of the TOL Plasmid of *Pseudomonas putida*." *J. Bacteriol.* 180: 2889–2894. 21

Marques, S. and J. L. Ramos (1993). "Transcriptional Control of the *Pseudomonas putida* TOL Plasmid Catabolic Pathways." *Mol. Microbiol.* 9(5): 923–929. DOI: 10.1111/j.1365-2958.1993.tb01222.x 21

Mermod, N., J. L. Ramos, et al. (1986). "Vector for regulated expression of cloned genes in a wide range of Gram-negative bacteria." *J. Bacteriol.* 167(2): 447–454.

Miller, W. G., J. H. Leveau, et al. (2000). "Improved *gfp* and *inaZ* broad-host-range promoter-probe vectors." *Mol. Plant. Microbe Interact.* 13(11): 1243–1250. DOI: 10.1094/MPMI.2000.13.11.1243 50, 52

Miller, W. G. and S. E. Lindow (1997). "An improved GFP cloning cassette designed for prokaryotic transcriptional fusions." *Gene* 191(2): 149–153. DOI: 10.1016/S0378-1119(97)00051-6 39

Minoia, M., M. Gaillard, et al. (2008). "Stochasticity and bistability in horizontal transfer control of a genomic island in *Pseudomonas*." *Proc. Natl. Acad. Sci. U. S. A.* 105(52): 20792–20797. DOI: 10.1073/pnas.0806164106

Mohn, W. W., J. Garmendia, et al. (2006). "Surveying biotransformations with a la carte genetic traps: translating dehydrochlorination of lindane (gamma-hexachlorocyclohexane) into *lacZ*-based phenotypes." *Environ. Microbiol.* 8(3): 546–555. DOI: 10.1111/j.1462-2920.2006.00983.x 41

Morett, E. and L. Segovia (1993). "The σ^{54} bacterial enhancer-binding protein family: Mechanism of action and phylogenetic relationship of their functional domains." *J. Bacteriol.* 175(19): 6067–6074. 40

Morgan-Kiss, R. M., C. Wadler, et al. (2002). "Long-term and homogeneous regulation of the *Escherichia coli araBAD* promoter by use of a lactose transporter of relaxed specificity." *Proc. Natl. Acad. Sci. U. S. A.* 99(11): 7373–7377. DOI: 10.1073/pnas.122227599 38

Ng, L. C., E. O'Neill, et al. (1996). "Genetic evidence for interdomain regulation of the phenol-responsive σ^{54}-dependent activator DmpR." *J. Biol. Chem.* 271(29): 17281–17286. DOI: 10.1074/jbc.271.29.17281 40

Norman, A., L. Hestbjerg Hansen, et al. (2005). "Construction of a ColD cda promoter-based SOS-green fluorescent protein whole-cell biosensor with higher sensitivity toward genotoxic compounds than constructs based on *recA, umuDC,* or *sulA* promoters." *Appl. Environ. Microbiol.* 71(5): 2338–2346. DOI: 10.1128/AEM.71.5.2338-2346.2005 32, 33

O'Neill, E., L. C. Ng, et al. (1998). "Aromatic ligand binding and intramolecular signalling of the phenol-responsive σ^{54}-dependent regulator DmpR." *Mol. Microbiol.* 28(1): 131–141. DOI: 10.1046/j.1365-2958.1998.00780.x 40

O'Neill, E., C. C. Sze, et al. (1999). "Novel effector control through modulation of a preexisting binding site of the aromatic-responsive σ^{54}-dependent regulator DmpR." *J. Biol. Chem.* 274(45): 32425–32432. DOI: 10.1074/jbc.274.45.32425 40

Paitan, Y., I. Biran, et al. (2004). "Monitoring aromatic hydrocarbons by whole cell electrochemical biosensors." *Anal. Biochem.* 335: 175–183. DOI: 10.1016/j.ab.2004.08.032 22

Park, S. J., J. Wireman, et al. (1992). "Genetic analysis of the Tn21 *mer* operator-promoter." *J. Bacteriol.* 174(7): 2160–2171. 26

Pedahzur, R., B. Polyak, et al. (2004). "Water toxicity detection by a panel of stress-responsive luminescent bacteria." *J. Appl. Toxicol.* 24: 343–348. DOI: 10.1002/jat.1023 31

Pedraza, J. M. and A. van Oudenaarden (2005). "Noise propagation in gene networks." *Science* 307: 1965–1969. DOI: 10.1126/science.1109090 46, 47

Pepi, M., D. Reniero, et al. (2005). "A comparison of MER::LUX whole cell biosensors and moss, a bioindicator, for estimating mercury pollution." *Water Air Soil Poll.* 173: 163–175. DOI: 10.1007/s11270-005-9043-4 26

Perez Martin, J., F. Rojo, et al. (1994). "Promoters responsive to DNA bending: A common theme in prokaryotic gene expression." *Microbiol. Rev.* 58(2): 268–290. 19, 36

Perez-Martin, J. and V. de Lorenzo (1995). "Integration host factor (IHF) suppresses promiscuous activation of the σ^{54}-dependent promoter Pu of *Pseudomonas putida*." *Proc. Natl. Acad. Sci. U.S.A.* 92: 7277–7281. DOI: 10.1073/pnas.92.16.7277 19

Platteeuw, C., G. Simons, et al. (1994). "Use of the *Escherichia coli* beta-glucuronidase (*gusA*) gene as a reporter gene for analyzing promoters in lactic acid bacteria." *Appl. Environ. Microbiol.* 60(2): 587–593.

Quillardet, P., O. Huisman, et al. (1982). "SOS chromotest, a direct assay of induction of an SOS function in *Escherichia coli* K-12 to measure genotoxicity." *Proc. Natl. Acad. Sci. U. S. A.* 79(19): 5971–5975. DOI: 10.1073/pnas.79.19.5971 52

Ramos, J. L., S. Marques, et al. (1997). "Transcriptional control of the *Pseudomonas* TOL plasmid catabolic operons is achieved through an interplay of host factors and plasmid- encoded regulators." *Annu. Rev. Microbiol.* 51: 341–373. DOI: 10.1146/annurev.micro.51.1.341 17, 19, 21

Reineke, W. and H.-J. Knackmuss (1988). "Microbial degradation of haloaromatics." *Annu. Rev. Microbiol.* 42: 263–287. DOI: 10.1146/annurev.mi.42.100188.001403 17

Rogowsky, P. M., T. J. Close, et al. (1987). "Regulation of the *vir* genes of *Agrobacterium tumefaciens* plasmid pTiC58." *J. Bacteriol.* 169(11): 5101–5112. 52

Rosen, B. P. (1995). "Resistance mechanisms to arsenicals and antimonials." *J. Basic Clin. Physiol. Pharmacol.* 6(3–4): 251-263. 28

Rosen, B. P. (1999). "Families of arsenic transporters." *Trends Microbiol.* 7(5): 207–212. DOI: 10.1016/S0966-842X(99)01494-8 28

Sarand, I., E. Skarfstad, et al. (2001). "Role of the DmpR-mediated regulatory circuit in bacterial biodegradation properties in methylphenol-amended soils." *Appl. Environ. Microbiol.* 67(1): 162–171. DOI: 10.1128/AEM.67.1.162-171.2001 41

Schell, M. A. (1993). "Molecular biology of the LysR family of transcriptional regulators." *Annu. Rev. Microbiol.* 47: 597–626. DOI: 10.1146/annurev.mi.47.100193.003121 17

Schell, M. A. and E. Poser (1989). "Demonstration, characterization, and mutational analysis of NahR protein binding in nah and sal promoters." *J. Bacteriol.* 171: 837–846. 17

Schweizer, H. P. (1991). "Improved broad-host-range *lac*-based plasmid vectors for the isolation and characterization of protein fusions in *Pseudomonas aeruginosa*." *Gene* 103(1): 87–92. DOI: 10.1016/0378-1119(91)90396-S

Schweizer, H. P. (1993). "Two plasmids, X1918 and Z1918, for easy recovery of the *xylE* and *lacZ* reporter genes." *Gene* 134(1): 89–91. DOI: 10.1016/0378-1119(93)90178-6

Scott, D. L., S. Ramanathan, et al. (1997). "Genetically engineered bacteria: electrochemical sensing systems for antimonite and arsenite." *Anal. Chem.* 69(1): 16–20. DOI: 10.1021/ac960788x 28

Selifonova, O., R. Burlage, et al. (1993). "Bioluminescent sensors for detection of bioavailable Hg(II) in the environment." *Appl. Environ. Microbiol.* 59(9): 3083–3090. 26

Shaner, N. C., R. E. Campbell, et al. (2004). "Improved monomeric red, orange and yellow fluorescent proteins derived from *Discosoma* sp. red fluorescent protein." *Nat. Biotechnol.* 22(12): 1567–1572. DOI: 10.1038/nbt1037 48, 49

Shaner, N. C., M. Z. Lin, et al. (2008). "Improving the photostability of bright monomeric orange and red fluorescent proteins." *Nat. Methods* 5(6): 545–551. DOI: 10.1038/nmeth.1209 48

Shingler, V. (1996). "Signal sensing by σ^{54}-dependent regulators: derepression as a control mechanism." *Mol. Microbiol.* 19(3): 409–416. DOI: 10.1046/j.1365-2958.1996.388920.x 19, 40

Shingler, V. and T. Moore (1994). "Sensing of aromatic compounds by the DmpR transcriptional activator of phenol-catabolizing *Pseudomonas* sp. strain CF600." *J. Bacteriol.* 176(6): 1555–1560. 19, 40

Shingleton, J. T., B. M. Applegate, et al. (1998). "Induction of the *tod* operon by trichloroethylene in *Pseudomonas putida* TVA8." *Appl. Environ. Microbiol.* 64(12): 5049–5052. 43

Silver, S. and T. Phung le (2005). "A bacterial view of the periodic table: genes and proteins for toxic inorganic ions." *J. Ind. Microbiol. Biotechnol.* 32(11–12): 587-605. DOI: 10.1007/s10295-005-0019-6 30, 53

Sinha, J., S. J. Reyes, et al. (2010) "Reprogramming bacteria to seek and destroy an herbicide." *Nat. Chem. Biol.*. 6:464–470. DOI: 10.1038/nchembio.369 42, 43

Skarfstad, E., E. O'Neill, et al. (2000). "Identification of an effector specificity subregion within the aromatic-responsive regulators DmpR and XylR by DNA shuffling." *J. Bacteriol.* 182: 3008–3016. DOI: 10.1128/JB.182.11.3008-3016.2000 41

Smirnova, I. A., C. Dian, et al. (2004). "Development of a bacterial biosensor for nitrotoluenes: the crystal structure of the transcriptional regulator DntR." *J. Mol. Biol.* 340(3): 405–418. DOI: 10.1016/j.jmb.2004.04.071 42

Stock, A. M., V. L. Robinson, et al. (2000). "Two-component signal transduction." *Annu. Rev. Biochem.* 69: 183–215. DOI: 10.1146/annurev.biochem.69.1.183 43, 44

Stocker, J., D. Balluch, et al. (2003). "Development of a set of simple bacterial biosensors for quantitative and rapid field measurements of arsenite and arsenate in potable water." *Environ. Sci. Technol.* 37: 4743–4750. DOI: 10.1021/es034258b 28, 30, 34, 35, 49

Summers, A. O. (1992). "Untwist and shout: a heavy metal-responsive transcriptional regulator." *J. Bacteriol.* 174(10): 3097–3101. 26

Tauriainen, S., M. Karp, et al. (1997). "Recombinant luminescent bacteria for measuring bioavailable arsenite and antimonite." *Appl. Environ. Microbiol.* 63(11): 4456–4461. 28

Tecon, R., O. Binggeli, et al. (2009). "Double-tagged fluorescent bacterial bioreporter for the study of polycyclic aromatic hydrocarbon diffusion and bioavailability." *Environ. Microbiol.* 11: 2271–2283. DOI: 10.1111/j.1462-2920.2009.01952.x 48, 49, 52

Tecon, R., M. Wells, et al. (2006). "A new green fluorescent protein-based bacterial biosensor for analysing phenanthrene fluxes." *Environ. Microbiol.* 8: 697–708. DOI: 10.1111/j.1462-2920.2005.00948.x 52

Tropel, D., A. B'ahler, et al. (2004). "Design of new promoters and of a dual-bioreporter based on cross-activation by the two regulatory proteins XylR and HbpR." *Environ. Microbiol.* 6(11): 1186–1196. DOI: 10.1111/j.1462-2920.2004.00645.x 24, 36

Tropel, D. and J. R. van der Meer (2002). "Identification and Physical characterization of the HbpR binding sites of the *hbpC* and *hbpD* promoters." *J. Bacteriol.* 184: 2914–2924. DOI: 10.1128/JB.184.11.2914-2924.2002 22

Tropel, D. and J. R. van der Meer (2004). "Bacterial transcriptional regulators for degradation pathways of aromatic compounds." *Microbiol. Mol. Biol. Rev.* 68(3): 474–500. DOI: 10.1128/MMBR.68.3.474-500.2004 14, 17, 19

Tropel, D. and J. R. van der Meer (2005). "Characterization of HbpR binding by site-directed mutagenesis of its DNA-binding site and by deletion of the effector domain." *FEBS J.* 272(7): 1756–1766. DOI: 10.1111/j.1742-4658.2005.04607.x 24

Turner, J. S., P. D. Glands, et al. (1996). "Zn2+-sensing by the cyanobacterial metallothionein repressor SmtB: different motifs mediate metal-induced protein-DNA dissociation." *Nucleic Acids Res.* 24(19): 3714–3721. DOI: 10.1093/nar/24.19.3714 30

Van Dyk, T. K., E. J. DeRose, et al. (2001). "LuxArray, a high-density, genomewide transcription analysis of *Escherichia coli* using bioluminescent reporter strains." *J. Bacteriol.* 183(19): 5496–5505. DOI: 10.1128/JB.183.19.5496-5505.2001 31

Virta, M., J. Lampinen, et al. (1995). "A luminescence-based mercury biosensor." *Anal. Chem.* 67(3): 667–669. DOI: 10.1021/ac00099a027 26

Vitale, E., A. Milani, et al. (2008). "Transcriptional wiring of the TOL plasmid regulatory network to its host involves the submission of the σ^{54}-promoter Pu to the response regulator PprA." *Mol. Microbiol.* 69(3): 698–713. DOI: 10.1111/j.1365-2958.2008.06321.x 19, 21, 35

Wackwitz, A., H. Harms, et al. (2008). "Internal arsenite bioassay calibration using multiple reporter cell lines." *Microb. Biotechnol.* 1(2): 149–157. DOI: 10.1111/j.1751-7915.2007.00011.x 30, 35, 49

Wade, J. T., N. B. Reppas, et al. (2005). "Genomic analysis of LexA binding reveals the permissive nature of the *Escherichia coli* genome and identifies unconventional target sites." *Genes Dev.* 19(21): 2619–2630. DOI: 10.1101/gad.1355605 31

Wells, M., M. Gösch, et al. (2005). "Ultrasensitive reporter protein detection in genetically engineered bacteria." *Anal. Chem.* 77: 2683–2689. DOI: 10.1021/ac048127k 49

Werlen, C., M. C. M. Jaspers, et al. (2004). "Measurement of biologically available naphthalene in gas, and aqueous phases by use of a *Pseudomonas putida* biosensor." *Appl. Environ. Microbiol.* 70: 43–51. DOI: 10.1128/AEM.70.1.43-51.2004 17

Wikstrom, P., E. O'Neill, et al. (2001). "The regulatory N-terminal region of the aromatic-responsive transcriptional activator DmpR constrains nucleotide-triggered multimerisation." *J. Mol. Biol.* 314(5): 971–984. DOI: 10.1006/jmbi.2000.5212 41

Willardson, B. M., J. F. Wilkins, et al. (1998). "Development and testing of a bacterial biosensor for toluene-based environmental contaminants." *Appl. Environ. Microbiol.* 64(3): 1006–1012. 21

Wise, A. A. and C. R. Kuske (2000). "Generation of novel bacterial regulatory proteins that detect priority pollutant phenols." *Appl. Environ. Microbiol.* 66(1): 163–169. DOI: 10.1128/AEM.66.1.163-169.2000 41

Wu, J. and B. P. Rosen (1993). "Metalloregulated expression of the *ars* operon." *J. Biol. Chem.* 268(1): 52–58. 28, 39

Yagur-Kroll, S., B. Bilic, et al. (2009). "Strategies for enhancing bioluminescent bacterial sensor performance by promoter region manipulation." *Microb. Biotechnol.* doi:10.1111/j.1751–7915.2009.00149.x. 32, 33

Ye, J., A. Kandegedara, et al. (2005). "Crystal structure of the *Staphylococcus aureus* pI258 CadC Cd(II)/Pb(II)/Zn(II)-responsive repressor." *J. Bacteriol.* 187(12): 4214–4221. DOI: 10.1128/JB.187.12.4214-4221.2005 28, 30

CHAPTER 3

Measuring with Bioreporters

3.1 ASSAY PRINCIPLES

3.1.1 RELATIVE AND ABSOLUTE MEASUREMENTS

Now that we have dealt with the concepts and details of the genetic circuitry in the cell that make up the sensing and reporting system, we can turn to the question of how to *assay* the reporter circuit in the sensor/reporter cells. In most cases, bioreporter assays are extremely simple; cells are maintained in some kind of buffered and equilibrated suspension and are brought into contact with a sample or calibrating solution for a particular time period, after or during which the reporter signal is recorded. In an 'inducible' reporter system (or 'lights on' – as it is sometimes called) one examines a *de novo* response of the cells in reaction to the target compound, which is higher than the background reporter signal. Incubation of the reporter cells in the assay even without specific target compound being present will result in some background reporter expression, the level of which is dependent on the type and characteristics of the components used for the sensor/reporter circuitry (as discussed in Chapter 2). The increase of reporter expression in the sample incubation containing the target compared to the reporter expression in the blank provides the basis for the measurement, as will be explained further below.

Some reporter designs rely on measuring a *decrease* in reporter signal (the 'lights-off' system). Dependent on the assay incubation time and type of reporter protein (stability, cofactors) the reporter signal decrease can be due to physiological factors that influence the reporter protein's activity or to a direct diminishment in reporter gene expression. The physiological influence on reporter activity is exploited in bioreporters which constitutively produce luciferase. Bioluminescence resulting from luciferase activity is a very energy- and cofactor-demanding process for the cell. Therefore, any interference with the energy status of the cell or with regeneration of the cofactors needed for luciferase activity will result in a decrease in light emission. A number of general toxicity bacterial sensors exist that operate in this manner, and also the Microtox®-system deploys this principle (see Chapter 1). Other designs use a combination of an inducible reporter (to detect a target compound) and a constitutive lights-off system (to report any toxicity interfering with the inducible response) [Biran et al., 2009].

Because of the nature of the bioreporter system and the assay layout, measurements of the reporter signal are in most cases *relative* measurements (Figure 3.1). An *absolute* measurement would consist of quantifying the exact number of reporter proteins produced in a cell, but in most assays with reporter cells an integrated specific activity of the reporter protein is measured over a large number of cells in the assay. An example of a relative measurement is the readout on a fluorimeter of

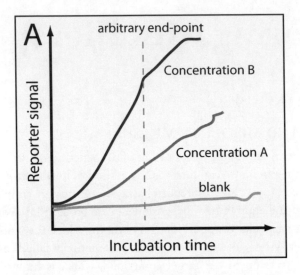

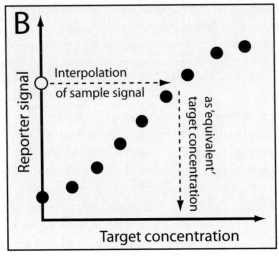

Figure 3.1: Bioassays with bacterial sensor/reporter cells in most cases produce *relative* measurements, because the reporter signal is dependent on the amount of cells in the assay and the incubation time (A). For this reason, bioassay measurements of target concentrations in unknown samples should be accompanied by a set of calibration assays with known and increasing target concentrations. The sample signal is then interpolated on a calibration curve to obtain the 'equivalent target concentration' in the sample (B).

a milliliter of assay suspension containing 10 million sensor cells each fluorescing due to a particular quantity of intracellularly produced gfp. The second reason of why bioreporter measurements are relative is because of the effect of incubation time on the induction process. When cells are left longer in contact with the target compound, the genetic circuitry may continue to elicit reporter gene expression and this will lead to more reporter protein in the cell, at least when the reporter is not an unstable protein and cells are not further dividing and diluting the signal. An integrated reporter signal measured in a bioreporter assay is thus a function of incubation time. Third reason for measurements being relative is the amount of cells in the assay. In practice it is very difficult to control the exact starting number of (bacterial) cells in a bioreporter assay. More cells on one day would automatically produce a higher integrated response than fewer cells in the assay on another day. Limited reproduction of the bacterial cells during the duration of the assay can also lead to a higher overall reporter signal, and also the growth phase of the cells (as explained in Chapter 2 for promoters experiencing exponential silencing) may influence the magnitude of induction.

In order to quantify the readout from a relative measurement on a set of samples one typically therefore uses a *calibration series* of simultaneous assays with known target concentrations. In this case, the assay readout for an unknown sample, which is incubated under exactly the same conditions as the calibration – same amount of reporter cells, same assay components and temperature, same incubation time – can be compared and interpolated to those of the calibration series to derive an *equivalent concentration* of target (Figure 3.1). What is the meaning of 'equivalent target concentration'? Remember that in most cases the sensing capacity of the regulatory protein used for the signal generation is not completely specific to only one chemical compound. For example, the ArsR-operator interaction is not only derepressed by arsenite, but also by antimonite with approximately similar efficiency [Wu and Rosen, 1993]. If the calibration series is carried out with arsenite then the response obtained in incubations with unknown samples should *au fond* be expressed as 'equivalent arsenite concentration', because in the unknown sample it might have been antimonite that elicits the bioreporter response. The use of equivalent concentrations often complicates bioreporter assay interpretations in the eyes of chemists, but if the bioassay is seen as a first line rapid analysis or when all targets are considered equally toxic or important, the use of equivalent concentrations is unproblematic. The fact of detecting the biological response of a complete mixture of compounds may actually form an additional advantage over chemical analytics of individual compounds.

3.1.2   END-POINT AND KINETIC MEASUREMENTS, MEASUREMENT TRANSFORMATIONS

A slight disadvantage of reporter systems involving *de novo* gene expression is, if one likes, the reaction time needed to develop the signal. This time is a combination of the time needed for the target compound to reach the cell, the time needed for gene transcription rate from the target switch and translation plus possibly posttranslational modifications of the reporter protein. In practice, the time needed for developing a sufficiently large reporter protein signal in an induction assay is between 30 minutes and 2 hours. This time depends on the type of reporter protein, whether the reporter protein's activity or presence can be assayed non-invasively or not, and on the assay and instrument configuration. In many cases one may settle for *end-point* measurements. This means that a fixed assay incubation time is chosen after which the reporter signal is read out, both in the standard calibration series and in the unknown samples. When the bioassays are very well standardized in terms of amount of cells, activation potential of the cells and incubation conditions, the responses are very reproducible and it may not be necessary to carry out full calibration series at every moment. The time of the end-point is dependent on the particularities of the sensor/reporter design and the host strains for the assay, but is generally chosen as such to have a maximum ratio of induction to background reporter signal within the shortest possible assay incubation time. As listed in Annexes 1-3, most published bacterial bioreporter assays use incubation times of 30 minutes to 2 hours but in exceptional cases or for specific research questions longer incubation times may be chosen.

As opposed to end-point measurements, one may also decide to make *kinetic* measurements. In this case, the regular increment of the reporter signal or in other words the *slope* of the regression line in the linear part of the reporter signal augmentation is measured[5] and plotted as a function of target compound concentration. Kinetic measurements are very suitable for reporter proteins of which the activity or abundance can be measured non-invasively and without destructing the cells (e.g., bacterial luciferase with intracellular substrate generation, or autofluorescent proteins). Furthermore, kinetic measurements can be useful for continuous on-line monitoring, for example, of process parameters, when the reporter cells can be maintained sufficiently well to ensure their proper and stable functioning.

On the basis of the primary measurements (end-point or kinetic), typically other secondary data treatments are performed. These include: background subtraction, logarithmic plotting, or calculation of induction factors (Figure 3.2). The advantage of background subtraction is that the data look nicer, although they are not more precise (see Section 3.5). Another way of presenting the data is to transform the primary reporter signals into induction factors that can then again be plotted as a function of target concentration. An induction factor is calculated by dividing the reporter signal at one particular concentration by that in the blank or by the signal at time zero. Induction factors are used to present the *magnitude* of the response to the target. When induction factors are very low, the sensor/reporter circuit is not very sensitive to the chemical target. Quantification will tend to be imprecise, because variations in repeated sample measurements will be more important than the increase in slope in the standard curve (see later Section 3.5). Under no circumstances should induction factors be calculated on background-subtracted data, because they will be artificially high. Finally, one often finds log-transformed data; for example, reporter output as function of the logarithm of the target concentration. Since most often the biological response follows a 'saturation' type, in which higher than optimal target concentrations do not produce a further increase of the response, the resulting calibration curves over a wide range of concentrations take the form of an S-shape. Regression line calculation can be performed on S-curves as for linear curves using residual minimization of the sum of the squares between the predicted and the effective measurements [Murk et al., 2002].

---

[5]In case the kinetics of reporter signal development cannot be described with a linear function, another fitting parameter is used.

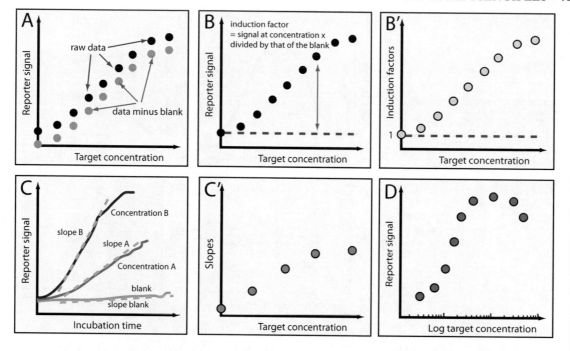

Figure 3.2: Various typical data treatments on biosensor/-reporter assays. (A) Background subtraction of the blank signal. (B) Normalization of reporter signal data into induction factors (B'), and a blank signal of 1. (C) Kinetic measurements of the slope of the signal increment at different concentrations and in the blank, transformed to a slopes-versus-target concentration plot (C'). (D) Log-transformed plots over wider concentration ranges. Signal decrease at higher than maximum is a result of toxicity.

Linear calibrations of biosensor/-reporter assays can be viewed as the response within a limited concentration range of the S-curve. Possibly, the signal at very high concentrations decreases again due to target toxicity (Figure 3.2 (D)).

3.1.3 POPULATION MEASUREMENTS

A number of concepts for measuring reporter signals are different from the general idea explained above, and have been applied for bioreporter assays in complex environments, in particular when information from individual reporter cells is acquired (e.g., *on planta*, in soil, or inside intestines, Figures 3.3, 3.4). Reporter expression from individual cells in a population is ideally Normal distributed around a mean (which one measures in a macro-assay across the whole population), however, in some instances expression in single cells is either very weak or not Normal distributed (Figure 3.3). Normal distribution can be tested by plotting cumulative data in normal probability plots, for which Normal distributed data appear as straight lines [Laemmli et al., 2000]. In cases of very weak re-

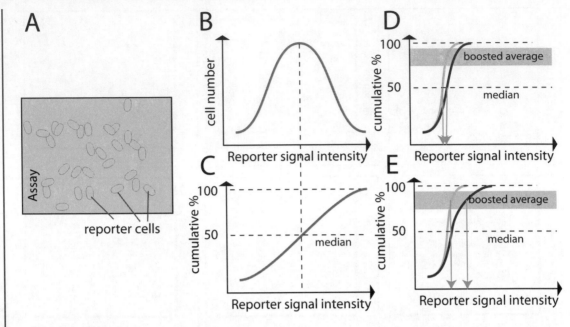

Figure 3.3: Reporter signal measurements and data transformation from individual cells. Digital image or flow cytometry reporter data from single cells (A) is represented as a regular distribution curve (B), from which a mean reporter signal intensity is calculated. Cumulative ranking of the same reporter signals across the population (C) provides an easy way to discern any aberrations from Normal distributions, which can be further visualized by presenting normal probability plots. In case when reporter signals between treatments differ very little or when the reporter signal distribution is not Normal, boosted averages (e.g., the average between the 70th and 95th percentile) can be used. Bootstrapping is used to calculate confidence intervals [Kohlmeier et al., 2007].

porter gene expression, we and others have calculated 'boosted averages', the average between the 70 and 95% percentile in a population of reporter cells cumulatively ranked according to the reporter intensity (Figure 3.3 (D)). The advantage of the boosted average is that differences in reporter signal between treatments are amplified, which makes their statistical discrimination more powerful [Kohlmeier et al., 2007]. Further more complex data treatments like neural network analysis can be used to identify the most optimally performing subpopulation of cells from cytometric data sets, such as developed by Wells and coworkers [Busam et al., 2007].

A further extreme case of population-dependent reporter expression is the one based on the toggle switch reporter explained in Section 2.8. This reporter – in theory - only produces no or maximal output, depending on the orientation of the switch of the promoter in front of the reporter gene. Because there is no gradual modulation of reporter signal in individual reporter cells,

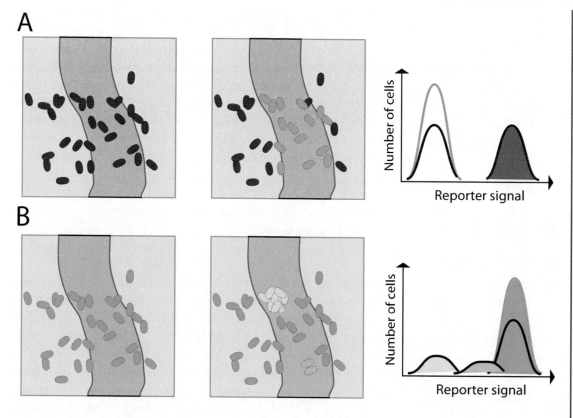

Figure 3.4: Population dependent bioreporter assays. (A) A toggle-switch bioreporter can only respond with no or maximum signal. The proportion of cells displaying the maximum signal after incubation into a particular complex environment (for example, a plant root) is thus a measure for the inducible response. (B) The reproductive history bioreporter is maximally induced at the start of the experiment but dilutes every time a cell divides. Rather than measuring new induction it measures the decrease of the reporter signal in a population of cells recovered after an incubation time in a particular environment. Reproduction of the cells can be derived from the proportions of cells with lower fluorescence.

the proportion of cells in a population displaying the maximum reporter signal is a more relevant measure for the inducible response [Casavant et al., 2003]. A useful application of this principle is that the information on the inducible response can be retrieved from basically any number of reporter cells (Figure 3.4). Reporter cells could thus be seeded in a complex environment and recovered after a certain incubation period, after which the response can be extracted from a small number of recovered reporter cells [Norman et al., 2006]. The second example is a form of 'lights-off' reporter, in which one starts with a population of reporter cells that all display the maximum signal of a stable reporter protein, but which dilutes every time the cells divide (Figure 3.4). The concept of the reporter is to measure the reproductive success of the reporter cells in the particular environment. The reporter circuit is based on the LacI repressed *lac* promoter, which becomes derepressed in the presence of an artificial effector compound like IPTG. The reporter cells are thus maximally derepressed at the start of the experiment and then introduced into a sampling environment of which one expects IPTG (or any other inducer for the *lac* operon) not to be present. Every time a cell divides the reporter signal will become twice as low (since there is no new induction); therefore, the reporter signal intensity compared to time zero and the number of cells displaying a particular status of reporter signal intensity can be interpreted in terms of the number of divisions that that particular reporter cell population has undergone over time [Remus-Emsermann and Leveau, 2010]. In practice, after four cell divisions the signal becomes too low to be detected. The system is a useful approach for quantifying, e.g., the amounts of available carbon or nutrients for cells in microenvironments like plant leafs, or in competition between different species.

3.1.4   SPIKING

Whereas calibration curves can be carried out under optimal conditions, measurements of real samples in the bioreporter assay may become problematic due to additional compounds being present in the sample, which may inhibit or stimulate the performance of the cells. In order to measure correctly, one therefore needs to adapt the medium composition for the calibration assay as close as possible to that of the assay including the sample. For example, if a seawater sample is to be measured in an otherwise saltless assay calibration medium, the calibration series has to include the introduced salt concentration from the seawater (or the sample has to be cleaned up to have the pure substance, see Section 3.4). Secondly, one may use so-called spiked measurements, in which a known concentration of the pure target is added to the sample. The measured increase in the spiked-sample can then be compared to the expected increase from the calibration curve, and, if the slopes of increase are different, can be used to estimate a correction factor (Figure 3.5). Use of spiking is recommended but correction factor estimation is not facile in the case of non-linear calibration curves. If the slope in the spiked control manifests too much inhibition, one should produce a dilution series of the sample (e.g., 0.3, 0.1, 0.03, 0.01 times) and repeat the measurement, or proceed with a specific sample cleanup before running the reporter assay. As a further alternative for correcting for possibly inhibitory substances in unknown samples, one could design a double reporter circuit as explained above, with one inducible parameter and one constitutive – the constitutive signal functioning as the inhibitory control measure [Biran et al., 2009].

3.2   THEORY OF ANALYTE PROVISION AND TRANSPORT

3.2.1   CALCULATION OF COMPOUND CONCENTRATIONS IN REPORTER CELLS

The reporter signal generated by the cells is thus dependent on the nature of the reporter circuit, the incubation time and the amount of target compound (analyte) experienced by the cells per unit

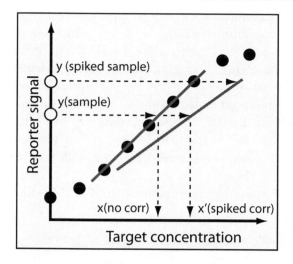

regular trend line of calibration:
$$y = ax + b$$
y = reporter signal
x = target concentration
a = slope (blue line)
b = intercept
$$x_{sample} = (y_{sample} - b)/a$$

spiked correction:
a' = slope of inhibition trend line =
$$\frac{(y_{spiked\ sample} - y_{sample})}{\text{spiked concentration difference}}$$
$$x_{sample,corr} = (y_{sample} - b)/a'$$

Figure 3.5: To control for possible inhibitory substances in the sample, a spiked assay is performed in which a known amount of pure target compound is added to the sample. When no inhibition is present, the augmentation of the reporter signal in the spiked sample will follow the slope calculated from the calibration trend line (in blue). If this is less then expected, an alternative trend line is calculated on the basis of the observed increase in the spiked versus the non-spiked sample (in red). This slope a' is then used to calculate the equivalent target concentration in the sample. Obviously, when a' becomes very flat a spiking correction can no longer be reasonably calculated and the sample should be diluted or cleaned up. The calculation example here assumes that the reporter output at zero target concentration is not changing in samples with an inhibitory composition to the bioreporter cells.

of time, and can be calibrated arbitrarily as explained above. On the contrary, it is not trivial to predict which amount of target compound passes through a cell per unit of time or is detected by the sensor/reporter circuit at any given external target compound concentration, although some relationship must exist between analyte concentration in the assay medium and the amount of analyte in the sensor/reporter cell [van der Meer et al., 2004]. To understand this relationship is important because the type of reporter assay (the 'macro-assay') and the nature of the designed reporter cell may influence the flux of target analyte to the cells, and thus determine the method of detection limit or detection sensitivity of the assay. As a first approach one could consider a typical bioreporter assay in which an arbitrary amount of cells is incubated in aqueous suspension (e.g., $5 \cdot 10^7$ cells per ml) with a soluble target compound (the sample) for a particular amount of time. We could further assume that this type of bacterial reporter has no specific means to actively transport target molecules from the bulk solution to their cell and into the cell's interior, and that all target molecules reach the reporter cells by molecular diffusion. Diffusive transport of soluble molecules is described by Fick's Laws, with an (overly simple but) easy approximation that the mean diffusion distance (x) is proportional to the square root of the diffusion time (t) times the molecule's diffusivity (D) times 4, from which follows that diffusion over twice the distance takes four times as long [Schwarzenbach et al., 1993]. At the typical diffusivities of dissolved compounds in water ($\sim 10^{-5}$ cm^2 s^{-1}), a mean cellular distance of ~ 36 μm in a suspension of $5{\cdot}10^7$ cells per ml is on average covered in roughly 80 msec. From this it follows that it is likely that contacts between all target molecules and all cells have occurred within the typical time of a biosensor assay of between 1 and 2 h. However, when diffusion is the only driving transport process and assuming that the cells' interior behaves like the bulk solution, under equilibrium the target compound concentration outside will equal that on the inside (see, for example, [Leveau et al., 1998]).

Assume a target compound with a dissolved aqueous phase concentration of 1 μM, which corresponds to

$$1 \cdot 10^{-6} \times 6.0 \cdot 10^{23} \text{ molecules per L} = 6.0 \cdot 10^{17} \text{ molecules per L, or } 6.0 \cdot 10^{14} \text{ per mL.}$$

At the assumed cellular density of $5 \cdot 10^7$ per mL in the assay, this would mean that:

$$6.0 \cdot 10^{14} \text{ divided by } 5 \cdot 10^7 = 1.2 \cdot 10^7 \text{ molecules}$$

would potentially be 'available' per cell and could contribute to eliciting the reporter signal. However, under the assumed equilibrium conditions of diffusive transport and no target compound accumulation in the reporter cells, and assuming further a typical bacterial cell volume of 1 fL (10^{-12} mL), only

$$6.0 \cdot 10^{14} \times 10^{-12} = 600 \text{ molecules}$$

would actually be present per sensor/reporter cell. These 600 molecules are obviously only a minute fraction of what was potentially available to the reporter cells at this concentration ($1.2{\cdot}10^7$). Equilibrium by diffusion can be expected to be reached in a few seconds [Leveau et al., 1998]. Also interesting to notice is that the fraction of compound inside the reporter cells (the *biota*) in the total assay would equal $600 \times 5{\cdot}10^7 = 3{\cdot}10^{10}$ molecules or less than 1 per mille. Interestingly, external target concentrations of 0.1-1 μM correspond roughly to the method of detection limit for many aqueous solution 'equilibrium' bioreporter assays (Annex 1-3), which is suggestive to conclude that some sort of physical transport limit is reached under such conditions. However, there is not sufficient experimental information at present to conclude whether a diffusion equilibrium is static (i.e., once 600 molecules in our example are reached, there is no further influx), or kinetic in the sense that many more inducer molecules pass into and out from the cells during the course of the bioassay, but that per unit of time only maximally 600 are present per cell. Furthermore, we have no current way of predicting whether the calculated 600 intracellular molecules can elicit gene expression only once (e.g., after which they stay bound to a regulator protein) or perhaps multiple times. But consider, for example, the published case of a mercury bioreporter with a method of detection limit of 1 fM [Virta et al., 1995]. At a concentration of 1 fM, the number of molecules per mL would become:

$$1 \cdot 10^{-15} \times 6.0 \cdot 10^{23} \text{ molecules per L} = 6.0 \cdot 10^{8} \text{ molecules per L, or} 6.0 \cdot 10^{5} \text{ per mL.}$$

At the assumed cell density of $5 \cdot 10^7$ per mL and further assuming equilibrium conditions, there would be clearly less than one molecule available per cell. This suggests that a single mercury molecule could recycle through many cells (via efflux and uptake), or trigger gene expression multiple times, leading to a remarkably sensitive reporter.

3.2.2 NON-DIFFUSIVE AND NON-CONSERVATIVE BIOREPORTERS

The bioreporter assay outlined above presents a relatively simple scenario, which in all likelihood is more complex in reality due to the presence of cellular or compound specific factors. Reporter cells may possess active uptake or efflux systems specific for the target compound, or be specifically engineered to include those in the cell (see Chapter 2). A simple diffusive transport equilibrium between exterior and the cell's interior can drastically change in the case of active uptake systems, which reportedly can accumulate some 100-fold with respect to the outside concentration (e.g., [Leveau et al., 1998]). In case of active uptake a 1 μM outside concentration would lead to a 100 μM inside concentration, which corresponds to $\sim$60,000 molecules per cell. At such active uptake the total fraction of target compound in the reporter cells would equal $5 \cdot 10^7$ cells $\times$ $6.0 \cdot 10^4$ molecules = $3 \cdot 10^{12}$, or $3 \cdot 10^{12}/6 \cdot 10^{14}$ = 0.5%. This is still a marginal fraction compared to the total amount of target compound in the assay and therefore unlikely to change further chemical equilibria (see below). We could thus conclude that potentially the inclusion of an active uptake system might increase the number of effector molecules eliciting a signal from the reporter circuit.

By contrast, reporter cells may also have efflux systems for the target compound or a modified target. This is particularly frequently the case with reporter cells engineered for heavy metal detection, because heavy metal detoxification systems often implicate efflux pumps [Silver and Phung le, 2005]. Efflux pump activity could potentially lower the intracellular effector concentration and, hence, increase the method of detection limit in a bioreporter assay. Indeed, experimental evidence from efflux mutants in various metal resistance systems support the idea that presence of an active efflux system increases and absence decreases the method of detection limit with roughly tenfold [Hynninen and Virta, 2010, Rensing et al., 2000, Stoyanov et al., 2003]. By contrast, so far the only experimental evidence on influence of an uptake system (from the mercury operon) suggested a meager twofold decrease of the method of detection limit in reporter strains with mercury transporter compared to without [Selifonova et al., 1993]. As an exception to the general effect of efflux systems it is worth to mention the *ars* defense system. In the natural *ars* system arsenite is pumped out of the cell by ArsBC efflux pump activity, but Tauriainen et al did not find any difference in method of detection limit for *E. coli* strains carrying an ars reporter circuit in the presence or absence of *arsBC* pump [Tauriainen et al., 1999]. This suggests that arsenite is removed from the cell by an as yet unknown other efflux system or that another mechanism may scavenge small amounts of arsenite. There is increasing evidence that this may actually be ArsR itself; apart from its regulatory role the protein also seems to act as scavenger protein for arsenite[6].

---

[6]See, for example recent iGEM competitions (http://2009.igem.org/Team:Groningen).

Further mechanisms may influence transport and partitioning processes of target compounds between cells and extracellular medium. In the abovementioned examples the total amount of target remains constant within the boundaries of the macro-assay and during the duration of the assay. One could imagine a micro-engineered system keeping reporter cells at one location but flowing sample over the cells. In this case, potentially many more molecules may pass through the cells and elicit signal. Other mechanisms may immobilize a target compound or change its chemical form, in order to withdraw it from the freely available target pool in the assay. For heavy metals, one could think of scavenging proteins (like ArsR or metallothioneins), of reductases (like the MerA Hg^{2+}-$Hg°$ reductase, leading to mercury volatilization), or of coprecipitation as insoluble salts (e.g., $PbPO_4$) [Hynninen et al., 2009]. As a consequence, of withdrawal of a chemical species from the assay, diffusion may continue leading to constant flux of target into the reporter cells (Figure 3.6).

In the case of organic and often lipophilic target molecules a partitioning will arise between the aqueous phase outside the cells, the lipophilic bacterial membranes, and the cytoplasmic inside (Figure 3.6). Depending on their lipophilic character, which is often represented by the octanol-water partition coefficient, compounds may become enriched 1000- to 100,000 fold in the membrane [Sikkema et al., 1995]. Assuming an approximate membrane thickness in a bacterial cell of some 20 nm, the membrane volume per bacterial cell would take up ~8%. In a partitioning equilibrium the bacterial membranes would thus by far carry the majority of such lipophilic compound as compared to the cytoplasm. Still, at most cell densities for biosensor measurements ($\sim5 \cdot 10^7$ cells per ml), the total mass fraction of compound in the bacterial cells will be very small ($< 5\%$). Interestingly, most regulatory proteins for organic compounds are still assumed to be cytoplasmic proteins [Tropel and van der Meer, 2004]. Despite a potentially much higher pool of target molecules in the membranes, such proteins would sample the (much lower) dissolved target compound concentration in the cytoplasm, but from an orthogonal reporter engineering point of view it would be much more interesting to have a regulatory protein detecting membrane dissolved target compound.

Even more than metals, organic compounds are amenable to transformation or degradation by the reporter cell, with the consequence that their disappearance will drive further target molecule transport to the reporter cells. As explained in Chapter 2 some reporter circuits even rely on transformation of an externally added target molecule to an internal metabolite that elicits the induction of the system. Prediction of intracellular effector concentrations at any given outside concentration becomes increasingly more difficult in the case of target compound metabolism, unless the kinetic parameters of compound uptake and metabolism are known. From our own work on a model system of 2,4-dichlorophenoxyacetic acid (2,4-D) degradation in *Cupriavidus* necator JMP134 we derived that 2,4-D diffusive uptake in the absence of active uptake system but in presence of induced metabolic pathway proceeded in the order of 110 molecules per sec per cell at an outside concentration of 1 μM [Leveau et al., 1998]. With the TfdK active uptake system and complete 2,4-D metabolism in place this rate increased to en estimated 2,100 molecules per sec per cell. At the $4 \cdot 10^8$ cells per mL used by the authors in their system this rate would implicate an overall

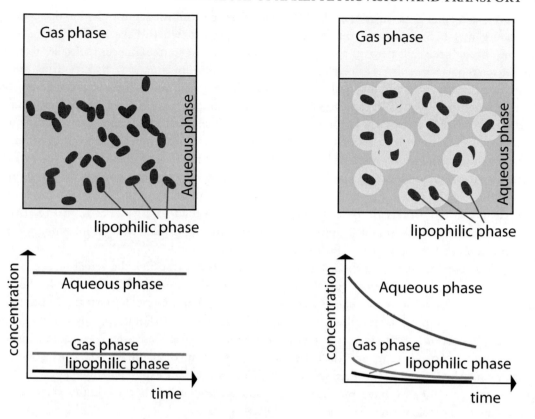

Figure 3.6: Destructive and non-destructive reporters. In the case of sensor cells that do not transform the target compound (left), the compound will be partitioned over gas phase, aqueous phase and lipophilic (or biomass) phase. The fraction of compound in each phase can be calculated according to chemical partitioning equilibrium, whereby the total amount of compound in the assay (assuming a closed vial) will not change. Destructive reporter cells will transform the target compound, which leads to its depletion over the course of the assay, but also to an increased flux of target compound towards the cells and through the reporter circuit. The total amount of target compound in the assay will diminish constantly and the phase partitioning is in a dynamic equilibrium.

$\sim 8{\cdot}10^{11}$ molecules disappearing per sec. To remove $6{\cdot}10^{14}$ molecules per mL (at 1 μM) would thus take roughly 1300 sec or 22 min, meaning that the assay would be completely depleted for 2,4-D within an assay time of 1 h. Hay et al reported a lower limit of detection for a *C. necator tfdC*-promoter based luciferase reporter for 2,4-D detection of 2 μM [Hay et al., 2000]. This reporter strain also degraded 2,4-D, suggesting that such a reporter would indeed need roughly 10^7 molecules per cell (as calculated earlier for 1 μM in equilibrium) to respond significantly above background.

However, as further data suggest, of these 10^7 molecules per cell the largest fraction disappears in compound metabolism. It would thus be interesting to speculate whether the 10^7 inducer molecules passing through the cells perhaps again only lead to the 600 or so free inducer molecules to twitch the sensor/reporter circuit (as calculated above for an equilibrium situation). By measuring mRNA formation directly in cells grown under continuous culture conditions, Leveau et al could also show that the magnitude of mRNA expression from the *tfd* promoters for 2,4-D degradation in *C. necator* JMP134 at 10 μM external 2,4-D concentration is only ephemerally detectable directly after a transition from medium without to medium with 2,4-D, but then decreases to levels non-significantly different from uninduced conditions [Leveau et al., 1999]. This 10 μM, to be correct, was added to a chemostat culture under steady state conditions, resulting in actual compound concentrations in the medium in the order of 10 nM, because of constant dilution of an incoming drop of medium (50 μl) into a 500 ml reactor. Indeed, Fuechslin et al measured 15 μg C per L or 0.15 μM actual concentration in chemostat cultures of *C. necator* growing on 2,4-D at an incoming medium concentration of 125 mg C per L [Füchslin et al., 2003].

This suggests that, once the 2,4-D metabolic pathway is formed, the actual inducer for the pathway (the metabolite 2,4-*cis,cis*-dichloromuconate, DCM) is more rapidly degraded further than that it can contribute to more gene expression from the *tfd* promoters. Reporters fused to one of the *tfd* promoters in the same cell will thus essentially not be able to 'intercept' any more signal. As discussed in Section 2.1, this is frustrating from the point of view of detecting 2,4-D as sensitively as possible, but it correctly represents the behavior of the bacterium degrading 2,4-D. Interestingly, the data obtained by Leveau also showed very clearly that when cells are exposed to higher outside 2,4-D concentrations (100 μM and 1 mM), *tfd* expression per se does not become higher but continues for a longer time, suggesting that the effector DCM is temporarily produced at a higher level than needed for RNA polymerase to process the *tfd* promoters at its maximum velocity [Leveau et al., 1999]. When DCM itself is transformed further and its intracellular concentration decreases to below the maximum threshold for gene expression, *tfd* expression goes down again [Leveau et al., 1999]. Reporter protein formation therefore does not follow the same pattern as mRNA expression. Similar conclusions can be drawn from a study of our group with a naphthalene reporter in *P. putida*, equipped with a *nahR-nahG'-luxAB* single copy reporter gene fusion and containing the NAH7 plasmid for naphthalene degradation [Werlen et al., 2004]. For bioreporter assays the cells were taken from a continuous culture on succinate and incubated in aqueous suspension with different naphthalene concentrations. We noticed that luciferase activity increased between 30 min and 3 h, but not for all naphthalene concentrations equally; after 30 min the reporter response was saturated at 0.5 μM naphthalene, after 1 h at 2 μM, and only after 2.5 h at 5 μM. Hence, only after 2.5 h a linear relation was obtained between luciferase activity and naphthalene concentration between 0 and 5 μM [Werlen et al., 2004]. This suggested that – like for the 2,4-D example – high concentrations of naphthalene temporarily lead to an overdose of the inducer metabolite salicylate in the cells, which takes time to become metabolized and prolongs induction of the reporter circuit. At longer incubation times lower naphthalene concentrations will be completely degraded but because of stability of the reporter protein its activity can still be measured after 2.5 h incubation. Further point of interest was that also the naphthalene reporter cells completely consumed the target compound during the assay. The method of detection limit of 0.5 μM (at $2.2 \cdot 10^7$ cells in the assay) indicated that $3 \cdot 10^{14}$ / $2.2 \cdot 10^7$ equals $\approx 10^7$ molecules per cell are necessary to obtain a reporter signal above background, which is similar to the values calculated above. Finally, this study demonstrated that the apparent method of detection limit in the assay can be lowered at least tenfold by changing the transport conditions of the target compound to the cells. With naphthalene this was possible by exploiting its volatility and allowing naphthalene to evaporate from a tenfold larger sample volume to be captured by the reporter cells exposed in the vapor phase on a filter [Werlen et al., 2004].

From these calculations, we can thus draw the conclusion that in most 'non-destructive' assays the number of inducer molecules is likely present in enormous overdose compared to the number of reporter cells and not all inducer molecules may contribute to eliciting the reporter circuit. Active uptake or deactivated efflux systems may increase the amount of target compound in the cell and decrease the method of detection limit in the bioreporter assay. Destructive reporter cells have the advantage of transforming the target compound and driving further compound diffusion to the cells, but the disadvantage that a large fraction of target compound is wasted in metabolism and is not contributing to reporter circuit induction.

3.3 CONCEPT OF BIOAVAILABILITY AND BIOACCESSIBILITY

3.3.1 BIOAVAILABLE FRACTIONS

From what we have discussed above we can see that the reporter cells 'sample' a fraction of target molecules that are at disposition in the total assay. Non-destructive reporter cells will likely not dramatically change the partitioning equilibrium between aqueous and vapor phase (at least at moderate cell densities; as explained above, less then 1% of compound is expected to be present in the biota). More lipophilic compounds will become more enriched in the cellular membrane but at low reporter cell densities presented above their fraction remains below 1% as well. Since target compound partitioning between cells/aqueous and vapor phase in a suspended cell aqueous mixture is a matter of seconds, one could thus define that the amount of target molecules that is present at any time during the assay in the reporter biomass (physically attached, dissolved in the membrane or dissolved in the cells' cytoplasm) composes the *bioavailable compound fraction* (Figure 3.7). According to the discussion above the bioavailable compound fraction would be different for a cell without any active transport ($\sim$600 molecules per cell at 1 μM outside concentration), with active transport ($\sim$60,000 molecules per cell at 1 μM), or with or without efflux. A destructive bioreporter cell would produce a very different bioavailable fraction (in the examples above $\sim 10^7$ molecules per cell at 1 μM outside concentration), and would not reach a partitioning equilibrium. In fact, after 30 minutes all compound may have been degraded. Hence, the signal that is reported by the cells under those assay conditions can be interpreted as a compound bioavailability measurement.

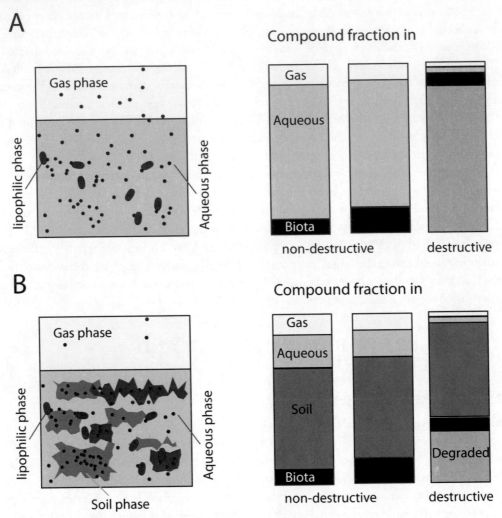

Figure 3.7: Bioavailable and bioaccessible fractions for different types of bioreporters. (A) Typical partitioning equilibrium in an aqueous phase assay with soluble target compound. A small fraction (not to scale) of compound partitions into reporter cell biomass (biota). This fraction may be different for the type of reporter cell, for example, one including an active uptake system or not. Destructive reporter cells will transform part of the target compound; possibly all during the time of the assay. The total compound fraction in the system diminishes. (B) More complex system in which target compound may be tightly bound to, e.g., soil particles. Non-destructive bioreporter cells will not significantly diminish the fraction of tightly bound compound and sample a partitioning equilibrium with aqueous and gas phases. Depending on the assay time, destructive bioreporter cells may retrieve part of the tightly bound fraction and sample the bioaccessible fraction.

The bioavailability concept is a very significant one in ecotoxicology, because it refers not to the total compound or dissolved concentration in the system, but to that fraction which is perceived by the biota and may exert its toxicity. Notwithstandingly, several recent discussions in scientific literature show that whereas bioavailability is a clearly definable notion, its quantitation and numerical value actually may depend on many factors, such as the type of compound, the organism, activity of that organism or time of contact [Harms and Bosma, 1997, Semple et al., 2007]. To quote from the most seminal paper on this subject, "bioavailability (is) defined as "that which is freely available to cross an organism's (cellular) membrane from the medium the organism inhabits at a given point in time" (…). Bioaccessibility (is) defined as "that which is available to cross an organisms' (cellular) membrane from the environment it inhabits, if the organism had access to it; however, it may be either physically removed from the organism or only bioavailable after a period of time" [Semple et al., 2007]. According to this definition, we can thus see how non-destructive reporter cells would measure bioavailability, because the freely available compound fraction will partition into the cell during the assay time, but the cells will not modify the compound. Destructive bioreporters will measure (different) bioavailable and bioaccessible fractions, depending on the time of the assay (Figure 3.7).

3.3.2   BIOAVAILABILITY AND BIOACCESSIBILITY REPORTER MEASUREMENTS

Performing reporter assay incubations of long duration such as needed for bioaccessibility measurements poses the difficulty that the behavior of the cells and formation of the reporter protein is less well controlled. One useful concept, which will be further illustrated below consists of using extremely stable reporter proteins such as gfp that record the history of exposure over longer time periods (e.g., days). This in itself bears the risk that the reporter cells will divide and dilute the signal on a *per cell* basis, in which case one would have to measure the total reporter protein amount in the population or use unstable reporters (as illustrated below). To substitute long assay incubation times several groups have experimented with non-exhaustive extraction procedures, in order to chemically extract the bioaccessible compound fraction from a complex sample and subsequently incubate this extract in a short duration bioreporter assay [Dawson et al., 2008, Paton et al., 2009]. We can illustrate bioreporter bioaccessibility measurements with two examples. In one case from our own research, we were interested to measure bioavailability and bioaccessibility of phenanthrene, a very poorly soluble polycyclic aromatic hydrocarbon. A reporter strain was constructed based on *Burkholderia sartisoli* RP007, a naphthalene and phenanthrene degrading bacterium. A plasmid reporter circuit was implanted into this bacterium consisting of the promoter for the gene *phnS* transcriptionally fused to a gene for stable egfp [Tecon et al., 2006]. Phenanthrene is so poorly soluble (< 7 mg per L) that no classical calibration curves with different compound concentrations in aqueous phase 2-3 h assays could be established, because the reporter signals were not significantly different from that in non-exposed cells. Reporter assays were thus developed to record the history of exposure over several days in the presence of phenanthrene as solid phase. Since the bioreporter degrades phenanthrene from solution it will drive further dissolution from the solid into the aqueous phase, which will result in a flux of phenanthrene 'through' the bioreporter cells, some of which (as in the example of the *P. putida* naphthalene reporter explained above) will elicit reporter gene expression. This phenanthrene flux is per definition a measure for the compound bioaccessibility to the reporter cells in that particular assay system. A sort of flux calibration could be established by using different solid phases of phenanthrene, e.g., crystals with different surface to volume ratio or materials onto which phenanthrene was sorbed. This resulted in measured fluxes of between 1 (for large crystals) and 60 fg phenanthrene per cell after 2 days (for small crystals); corresponding to $1 \cdot 10^{-15}$ / 182 = $5 \cdot 10^{-18}$ mol, or $5 \cdot 10^{-18} \times 6 \cdot 10^{23} = 3 \cdot 10^{6}$ molecules per cell at the lowest response; a value not unlike the 10^{7} molecules per cell necessary to elicit response from the naphthalene bioreporter in a 3 h assay. Whereas these assays still consisted of reporter cells incubated in aqueous phase with the solid phase phenanthrene, our group also experimented with diffusion-based assays, in which cells are embedded in agarose gel and exposed to a point source from which the target compound diffuses over time of the assay (Figure 3.8). The advantage of this assay is that the availability of compound from a complex source can be observed over time without changing the integrity of the assay [Tecon et al., 2009]. To compensate for growth of the reporter cells (and consequently the decrease of the reporter signal) one can include a second constitutively produced marker in the cells to be used as a cell counter.

In the second example Leveau et al studied availability of fructose on plant leaves [Leveau and Lindow, 2001]. In contrast to the foregoing reporters they developed a reporter with an unstable egfp in order not to record the history of exposure but rather to detect at any time the active population of reporter cells. Their *Erwinia herbicola fruBp-egfp(AAV)* reporter was sprayed on bean leaves and cells were washed from leaves at different time intervals to record the egfp signal intensity and the proportion of cells with egfp. Interestingly, whereas most cells produced egfp after short time incubation (1, 3 or 7 h), only very few still produced egfp after 24 h (see, e.g., Figure 3.3). This meant that most cells had actually metabolized the available sugar within a few hours, but some cells had apparently been exposed to pockets of more available sugar on the leaf from which they could profit even after 24 h. Estimations from local growth rates and sugar consumption suggested that ~84% of the initially applied cells are exposed to 0.15 pg, which does not permit a single cell doubling. Some 11% of the population of cells applied to leaves had access to 4.6 pg, allowing a few cell doublings (Leveau et al assumed 0.4 pg fructose needed for one cell division). An estimated quantity of 4.4 ng fructose was available to 1 to 2 percent of cells, those which showed egfp even after 24 h [Leveau and Lindow, 2001].

3.4   BIOREPORTER ASSAY TYPES

3.4.1   AQUEOUS ASSAYS

The simplest form of a bioassay with bacterial reporter cells is the one where the cells are suspended in aqueous solution with the sample, incubated for the appropriate amount of time, after which the reporter signal is recorded (Figure 3.8). In case of continuous measurement of the reporter signal, the assay can be incubated for as long as is necessary to obtain a significantly different response from the

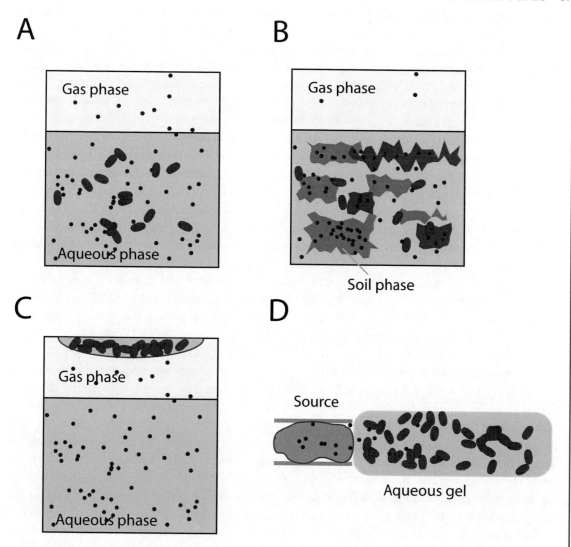

Figure 3.8: Bioreporter assay types. (A) Aqueous suspension. Reporter expression can be measured in individual cells or from the population as a whole. (B) Aqueous suspension including solid phase. Reporter expression is typically measured on cells recovered from the assay to avoid diffraction of (light) signal on solid phase particles. (C) Gas phase assay for volatile target compounds. Cells are incubated in a hanging droplet or on a filter on top of a gel to avoid desiccation. Assay can be performed in presence of solid phase. (D) Source-diffusion assay. Source with target compound is brought into contact with reporter cells immobilized in a gel. Diffusion from source will result in a target compound gradient being formed over time. Reporter expression is measured as a function of time and distance to the source.

blank (see Figure 3.1). Simple as these bioreporter assays are, a number of considerations will need to be made with respect to buffer composition, presence or not of growth substrates, precultivation and growth state of the cells, closed or open assays. With the growth state, microbiologists generally refer to 'exponentially growing' versus 'non-growing' or stationary phase cells. Exponentially growing is really a somewhat bad terminology, since it is often taken as an equivalent to the phase of growth in a batch culture when all cells are reproducing at maximum rate (in that particular growth medium). It would, however, probably be more correct to refer to the exact growth rate during time of the precultivation or in the assay, but this is more difficult to achieve. What is the most important for the assay is that none of the parameters in the assay medium is inhibiting the cells from inducing or derepressing the promoter controlling the reporter gene in the presence of the target compound.

In practice, therefore, the bacterial reporter cells are cultivated on a general growth medium to obtain the necessary cell numbers for the assay(s), counting all triplicates, sample numbers and calibration series, harvested during the phase when they are all reproducing exponentially, and diluted in the assay medium of the same composition as the growth medium – except perhaps for presence of the growth substrate. The idea in this procedure is to have the most 'active' cells, which react immediately to the target compound. Many reporter cells can also simply be frozen in growth medium plus 15% glycerol during exponential phase and stored at -80°C until used for the assay [Tecon et al., 2010]. Long term storage of active reporter cells in freeze-dried form is not altogether trivial but many mysteries remain about the optimal protocols [Bjerketorp et al., 2006]. Frozen aliquots can be removed for the assay, preincubated for a couple of minutes at 20°C, after which they will be immediately reacting to target conditions (in our experience, microscopic imaging of gfp reporter cells showed that actually some 20% of the cells may lyse or form ghost cells upon thawing frozen aliquots). In other cases reporter cells may have to be grown in continuous culture in order to obtain the most rapidly reacting cells [Werlen et al., 2004]. Toxicity reporter strains may have to be grown with care (i.e., no changes in growth temperature or medium) or else they display higher background signals due to induction of repair programs whilst growing [Norman et al., 2005]. Heavy metal responsive reporter strains can usually be grown and assayed in any type of medium (except for the metals under consideration), since the reporter circuits are not influenced by the carbon or nitrogen substrates. In contrast, reporter strains responsive to organic compound may be sensitive to catabolite repression and care should be taken to provide an energy source to allow reporter protein synthesis but which should not inhibit induction of the reporter circuit [Marques et al., 2006].

The chosen reporter cell density, as discussed above, may be dependent on the type and sensitivity of the instrument needed to record the reporter signal from the cells. We find that for all-round bioassays with bacterial reporter strains based on *E. coli, P. putida, Burkholderia, P. fluorescens,* etc., a cell density of between 2 and $5 \cdot 10^7$ cells per ml is by far sufficient, with luciferase, gfp or beta-galactosidase as reporter.

The matrix of the aqueous sample itself is not always unproblematic for the reporter assay, due to the presence of other contaminating and inhibitory compounds (e.g., heavy metals and organics), the presence of potentially inhibitory proteins or peptides (e.g., urine or blood samples), or the salt concentration (e.g., marine samples). We find that most *E. coli*-based reporter strains can function in marine samples when these are diluted at least four times (and the calibration series is carried out in the same diluted marine water) [Tecon et al., 2010]. Others showed that an *E. coli* bioreporter even functioned flawless in diluted blood serum [Turner et al., 2007] or diluted (crab) urine [Lewis et al., 2009]. When dilution is compromising the compound detection limit in the assay one would have to extract the sample and clean it up before performing the assay. Most reporter cells, however, are sensitive to organic solvents, although low concentrations of dimethylsulfoxide or methanol (below 1%) may be tolerated [Patterson et al., 2004] [Paton et al., 2009].

3.4.2   GAS PHASE MEASUREMENTS

Reporter measurements in the gas or vapor phase may have a number of advantages over aqueous phase assays with suspended cells. Many toxic organic compounds may be sufficiently volatile that they will evaporate from a sample. In this case, measuring in the gas phase has the advantage that compound diffusivities are much faster than in aqueous phase, which, as demonstrated above, can be exploited to 'concentrate' target compound from larger sample sizes on smaller volumes of reporter cells. This is useful because lower method of detection limits can be achieved [Werlen et al., 2004]. An additional advantage of sampling the gas phase with the reporter cells is that potentially toxic non-volatile compounds in the sample will not disturb the assay. For example, toxic heavy metals are often present in the same contaminated sample as organic compounds and may lead to underperformance of the reporter cells. However, since they are not volatile they will not inhibit reporter cells that are incubated in the vapor phase. Third advantage of assaying in the gas phase is that complex samples (e.g., soil slurries) can be incubated without any risk for light diffraction of the reporter signal from particulate matter [Kohlmeier et al., 2008].

In most described gas phase assays the reporter cells are not actually incubated in air dry state, but typically exposed as concentrated cell suspensions in small drops with aqueous medium hanging at the bottom of the lid closing the sampling vial [Deepthike et al., 2009, Kohlmeier et al., 2008], on filters on top of agar medium in small tubes within sampling vials to prevent drying out [Werlen et al., 2004], or on agar plates in the vicinity of a volatile source [de las Heras et al., 2008]. Reporter cells have also been immobilized in gels on tips of optical fibers [Heitzer et al., 1994], which permits to assay the reporter response in the vapor phase. Future interesting approaches may consist of sampling atmospheric contamination via reporter cells attached to leaf surfaces [Sandhu et al., 2007]. Anecdotal evidence suggests that the ubiquitous presence of oxygen around small droplets of reporter cells hanging in the gas phase is advantageous for maturation and activity of reporters such as gfp or luciferase, but this has not been investigated systematically in comparison to liquid assays.

3.4.3   SOLID PHASE ASSAYS

For many purposes assays including solid phases such as soils or sediments would be very useful. The reason for this is that the source of contamination is very often the solid material itself (e.g., soil, sediment), and sampling the groundwater is not doing justice to assess the level of contamination at a site. Other tests may assay solid food stuffs, or sample complex materials like tissues. As far as I am aware no bioreporter assay directly functions with solid phase to reporter cell contact, but always via an intermediate of (some) aqueous or gas phase. Various procedures have thus been developed to extract sorbed chemicals from solid phases or enzymatically destroy a food matrix and release the target chemicals for the bioreporter assay. For example, non-exhaustive extraction methods were used to retrieve different naphthalene fractions from contaminated soils and correlate the bioreporter responses to the compound fraction mineralized by microbial activity in such soil [Paton et al., 2009, Patterson et al., 2004]. Of various chemicals tested (e.g., cyclodextrins, methanol, water, XAD-4), a hydroxypropyl-beta-cyclodextrin extract produced a bioreporter response that correlated the closest with mineralization rates, although methanol retrieved the largest compound fraction from the soils and sediments [Paton et al., 2009]. In principle also biosurfactants can be used for target chemical extraction from contaminated solids [Tecon and van der Meer, 2009]. In our own work we developed an effective procedure to extract inorganic arsenicals from contaminated rice, using finely grounded powder and enzymatic digestion with pancreatin at low pH and a temperature of 55°C [Baumann and van der Meer, 2007]. Although such procedures prolong the bioreporter assay they are necessary in order to obtain faithful signals and meaningful values. Bioreporter assays were also developed to quantify fructose availability on plant leaves [Leveau and Lindow, 2001], tetracycline in pig feces [Hansen et al., 2002], N-acylhomoserine lactones in tissue or soil [Burmølle et al., 2003], or tetracyclines in (homogenized) fish tissue [Pellinen et al., 2002]. In the case of bioreporter assays on plant leaves, roots, intestines or soil, the reporter cells were incubated with the sample and then specifically separated from the sample matrix in order to avoid reporter signal bleaching or diffraction. Separation of reporter cells from soil particles can be carried out using Nycodenz gradients, after which the reporter signal in individual cells can be analyzed using flow cytometry [Burmølle et al., 2003]. Alternatively, reporter cells can be incubated with the soil or sediment but remain in a separated compartment such as a dialysis chamber, which can be removed after the required contact time and analyzed for reporter expression [Deepthike et al., 2009]. In case of volatile substances – as described above, the reporter cells can also be incubated at a distance from a soil or sediment sample in a hanging droplet or on a filter [Kohlmeier et al., 2008, Werlen et al., 2004].

Finally, we also experimented with embedding reporter cells in small agarose strips, which are brought into contact to solid materials on one end. Through dissolution and molecular diffusion a gradient of the target compound is formed in the agarose strip, which will change over time and as a function of activity of the reporter cells (Figure 3.8). By analyzing the reporter expression across distance and at different incubation periods, a cumulative reporter signal can be obtained representing the total accessible compound fraction in that material [Tecon et al., 2009].

3.5   METHOD DETECTION LIMITS, ACCURACY AND PRECISION

A final word is needed on the accuracy of bioreporter assays. Typically, one would wish to calculate figures of merit, which include the method of detection limit for the analyte (MDL), average relative standard deviation based upon replicate measurements (RSD), the uncertainty associated with the determination of an unknown sample using a calibration curve, the correlation coefficient of the calibration curve (r^2) and the calibration sensitivity (S). The MDL can be calculated in a similar manner as for chemical analysis as being the calculated analyte concentration corresponding to an ordinate value of the blank plus three times the standard deviation of measurements on the blank, using the calibration curve [Wells et al., 2005]. Other texts use the following definitions for the MDL: (i) the concentration that causes a fold-induction of $(1 + 2 \cdot CVb)/(1 - 2 \cdot CVb)$ compared to the blank, with CVb being the coefficient of variation in the blank measurements [Hynninen and Virta, 2010]; (ii) a concentration that causes a fold-induction of $2 + (6 \cdot s_b/S_b)$ compared to the blank, with s_b being the standard deviation in the blank measurements and S_b being the signal in the blank [Ivask et al., 2009]. For an overview of reported MDLs see Annex 1 and 2. It should be pointed out the the MDL is not only a characteristic of the reporter strain or cell line, but mostly of the assay measurements, numbers of replicates and instruments [Wells et al., 2005]. The measurement precision is equivalent to the average relative standard deviation over all replicates, which again depends on the number of replicates and the instrument use for quantifying the bioreporter response. Not unsurprisingly, for example, epifluorescence microscopy is less precise in quantifying egfp signals than is steady state fluorimetry [Kohlmeier et al., 2007, Wells et al., 2005]. Sensitivity of the assay refers to the percent increase of the signal over time or as a function of analyte concentration.

REFERENCES

Baumann, B. and J. R. van der Meer (2007). "Analysis of bioavailable arsenic in rice with whole cell living bioreporter bacteria." *J. Agric. Food Chem.* 55(6): 2115–2120. DOI: 10.1021/jf0631676 92

Biran, A., R. Pedahzur, et al. (2009). Genetically engineered bacteria for genotoxicity assessment. *Biosensors for the Environmental Monitoring of Aquatic Systems*. D. Barcelo and P.-D. Hansen. Berlin/ Heidelberg, Springer: 161–186. 71, 78

Bjerketorp, J., S. Hakansson, et al. (2006). "Advances in preservation methods: keeping biosensor microorganisms alive and active." *Curr. Opin. Biotechnol.* 17(1): 43–49. DOI: 10.1016/j.copbio.2005.12.005 90

Burmølle, M., L. H. Hansen, et al. (2003). "Presence of N-Acyl Homoserine Lactones in Soil Detected by a Whole-Cell Biosensor and Flow Cytometry." *Microb. Ecol.* 45: 226–236. DOI: 10.1007/s00248-002-2028-6 92

Busam, S., M. McNabb, et al. (2007). "Artificial neural network study of whole-cell bacterial biosensor response determined using fluorescence flow cytometry." *Anal. Chem.* 79: 9107–9114. DOI: 10.1021/ac0713508 76

Casavant, N. C., D. Thompson, et al. (2003). "Use of a site-specific recombination-based biosensor for detecting bioavailable toluene and related compounds on roots." *Environ. Microbiol.* 5: 238–249. DOI: 10.1046/j.1462-2920.2003.00420.x 77

Dawson, J. J., C. O. Iroegbu, et al. (2008). "Application of luminescent biosensors for monitoring the degradation and toxicity of BTEX compounds in soils." *J. Appl. Microbiol.* 104(1): 141–151. DOI: 10.1111/j.1365-2672.2007.03552.x 87

de las Heras, A., C. A. Carreno, et al. (2008). "Stable implantation of orthogonal sensor circuits in Gram-negative bacteria for environmental release." *Environ. Microbiol.* 10(12): 3305–3316. DOI: 10.1111/j.1462-2920.2008.01722.x 91

Deepthike, H. U., R. Tecon, et al. (2009). "Unlike PAHs from Exxon Valdez crude oil, PAHs from Gulf of Alaska coals are not readily bioavailable." *Environ. Sci. Technol.* 43: 5864–5870. DOI: 10.1021/es900734k 91, 92

Füchslin, H.-P., I. Rüegg, et al. (2003). "Effect of integration of a GFP reporter gene in *Ralstonia eutropha* on growth kinetics with 2,4-dichlorophenoxyacetic acid." *Environ. Microbiol.* 5(10): 878–887. DOI: 10.1046/j.1462-2920.2003.00479.x 84

Hansen, L. H., F. Aarestrup, et al. (2002). "Quantification of bioavailable chlortetracycline in pig feces using a bacterial whole-cell biosensor." *Vet. Microbiol.* 87(1): 51–57. DOI: 10.1016/S0378-1135(02)00029-9 92

Harms, H. and T. N. P. Bosma (1997). "Mass transfer limitation of microbial growth and pollutant degradation." *J. Ind. Microbiol. Biotechnol.* 18: 97–105. DOI: 10.1038/sj.jim.2900259 87

Hay, A. G., J. F. Rice, et al. (2000). "A bioluminescent whole-cell reporter for detection of 2,4-dichlorophenoxyacetic acid and 2,4-dichlorophenol in soil." *Appl. Environ. Microbiol.* 66: 4589–4594. DOI: 10.1128/AEM.66.10.4589-4594.2000 83

Heitzer, A., K. Malachowsky, et al. (1994). "Optical biosensor for environmental on-line monitoring of naphthalene and salicylate bioavailability with an immobilized bioluminescent catabolic reporter bacterium." *Appl. Environ. Microbiol.* 60(5): 1487–1494. 91

Hynninen, A., T. Touze, et al. (2009). "An efflux transporter PbrA and a phosphatase PbrB cooperate in a lead-resistance mechanism in bacteria." *Mol. Microbiol.* 74(2): 384–394. DOI: 10.1111/j.1365-2958.2009.06868.x 82

Hynninen, A. and M. Virta (2010). "Whole-cell bioreporters for the detection of bioavailable metals." *Adv. Biochem. Eng. Biotechnol.* 118: 31–63. DOI: 10.1007/10_2009_9 81, 93

Ivask, A., T. Rolova, et al. (2009). "A suite of recombinant luminescent bacterial strains for the quantification of bioavailable heavy metals and toxicity testing." *BMC Biotechnol.* 9: 41. DOI: 10.1186/1472-6750-9-41 93

Kohlmeier, S., M. Mancuso, et al. (2008). "Comparison of naphthalene bioavailability determined by whole-cell biosensing and availability determined by extraction with Tenax." *Environ. Pollut.* 156(3): 803–808. DOI: 10.1016/j.envpol.2008.06.001 91, 92

Kohlmeier, S., M. Mancuso, et al. (2007). "Bioreporters: gfp versus lux revisited and single-cell response." *Biosens. Bioelectron.* 22(8): 1578–1585. DOI: 10.1016/j.bios.2006.07.005 76, 93

Laemmli, C. M., J. H. J. Leveau, et al. (2000). "Characterization of a second *tfd* gene cluster for chlorophenol and chlorocatechol metabolism on plasmid pJP4 in *Ralstonia eutropha* JMP134(pJP4)." *J. Bacteriol.* 182(15): 4165–4172. DOI: 10.1128/JB.182.15.4165-4172.2000 75

Leveau, J. H. and S. E. Lindow (2001). "Appetite of an epiphyte: Quantitative monitoring of bacterial sugar consumption in the phyllosphere." *Proc. Natl. Acad. Sci. U. S. A.* 98(6): 3446–3453. DOI: 10.1073/pnas.061629598 88, 92

Leveau, J. H., A. J. B. Zehnder, et al. (1998). "The *tfdK* gene product facilitates uptake of 2,4-dichlorophenoxyacetate by *Ralstonia eutropha* JMP134(pJP4)." *J. Bacteriol.* 180(8): 2237–2243. 80, 81, 82

Leveau, J. H. J., F. König, et al. (1999). "Dynamics of multigene expression during catabolic adaptation of *Ralstonia eutropha* JMP134 (pJP4) to the herbicide 2, 4-dichlorophenoxyacetate." *Mol. Microbiol.* 33(2): 396–406. DOI: 10.1046/j.1365-2958.1999.01483.x 84

Lewis, C., S. Beggah, et al. (2009). "Novel use of a whole cell *E. coli* bioreporter as a urinary exposure biomarker." *Environ. Sci. Technol.* 43(2): 423–428. DOI: 10.1021/es801325u 91

Marques, S., I. Aranda-Olmedo, et al. (2006). "Controlling bacterial physiology for optimal expression of gene reporter constructs." *Curr. Opin. Biotechnol.* 17(1): 50–56. DOI: 10.1016/j.copbio.2005.12.001 90

Murk, A. J., J. Legler, et al. (2002). "Detection of estrogenic potency in wastewater and surface water with three in vitro bioassays." *Environ. Toxicol. Chem.* 21(1): 16–23. DOI: 10.1002/etc.5620210103 74

Norman, A., L. H. Hansen, et al. (2006). "A flow cytometry-optimized assay using an SOS-green fluorescent protein (SOS-GFP) whole-cell biosensor for the detection of genotoxins in complex environments." *Mutat. Res.*(603). 164–173. DOI: 10.1016/j.mrgentox.2005.11.008 77

Norman, A., L. Hestbjerg Hansen, et al. (2005). "Construction of a ColD cda promoter-based SOS-green fluorescent protein whole-cell biosensor with higher sensitivity toward genotoxic compounds than constructs based on *recA, umuDC*, or *sulA* promoters." *Appl. Environ. Microbiol.* 71(5): 2338–2346. DOI: 10.1128/AEM.71.5.2338-2346.2005 90

Paton, G. I., B. J. Reid, et al. (2009). "Application of a luminescence-based biosensor for assessing naphthalene biodegradation in soils from a manufactured gas plant." *Environ. Pollut.* 157(5): 1643–1648. DOI: 10.1016/j.envpol.2008.12.020 87, 91, 92

Patterson, C. J., K. T. Semple, et al. (2004). "Non-exhaustive extraction techniques (NEETs) for the prediction of naphthalene mineralisation in soil." *FEMS Microbiol. Lett.* 241(2): 215–220. DOI: 10.1016/j.femsle.2004.10.023 91, 92

Pellinen, T., G.-R. Bylund, et al. (2002). "Detection of Traces of Tetracyclines from Fish with a Bioluminescent Sensor Strain Incorporating Bacterial Luciferase Reporter Genes." *J. Agric. Food Chem.* 50: 4812–4815. DOI: 10.1021/jf0204021 92

Remus-Emsermann, M. N. and J. H. Leveau (2010). "Linking environmental heterogeneity and reproductive success at single-cell resolution." *ISME J.* 4(2): 215–222. DOI: 10.1038/ismej.2009.110 78

Rensing, C., B. Fan, et al. (2000). "CopA: An *Escherichia coli* Cu(I)-translocating P-type ATPase." *Proc. Natl. Acad. Sci. U. S. A.* 97(2): 652–656. DOI: 10.1073/pnas.97.2.652 81

Sandhu, A., L. J. Halverson, et al. (2007). "Bacterial degradation of airborne phenol in the phyllosphere." *Environ. Microbiol.* 9(2): 383–392. DOI: 10.1111/j.1462-2920.2006.01149.x 91

Schwarzenbach, R. P., P. M. Gschwend, et al. (1993). *Environmental organic chemistry.* New York, John Wiley & Sons, Inc. 80

Selifonova, O., R. Burlage, et al. (1993). "Bioluminescent sensors for detection of bioavailable Hg(II) in the environment." *Appl. Environ. Microbiol.* 59(9): 3083–3090. 81

Semple, K. T., K. J. Doick, et al. (2007). "Microbial interactions with organic contaminants in soil: Definitions, processes and measurement." *Environ. Poll.*(150). DOI: 10.1016/j.envpol.2007.07.023 87

Sikkema, J., J. A. de Bont, et al. (1995). "Mechanisms of membrane toxicity of hydrocarbons." *Microbiol. Rev.* 59(2): 201–222. 82

Silver, S. and T. Phung le (2005). "A bacterial view of the periodic table: genes and proteins for toxic inorganic ions." *J. Ind. Microbiol. Biotechnol.* 32(11–12): 587-605. DOI: 10.1007/s10295-005-0019-6 81

Stoyanov, J. V., D. Magnani, et al. (2003). "Measurement of cytoplasmic copper, silver, and gold with a lux biosensor shows copper and silver, but not gold, efflux by the CopA ATPase of *Escherichia coli.*" *FEBS Lett.* 546(2–3): 391-394. DOI: 10.1016/S0014-5793(03)00640-9 81

Tauriainen, S., M. Virta, et al. (1999). "Measurement of firefly luciferase reporter gene activity from cells and lysates using *Escherichia coli* arsenite and mercury sensors." *Anal. Biochem.* 272: 191–198. DOI: 10.1006/abio.1999.4193 81

Tecon, R., S. Beggah, et al. (2010). "Development of a Multistrain Bacterial Bioreporter Platform for the Monitoring of Hydrocarbon Contaminants in Marine Environments." *Environ. Sci. Technol.* 144: 1049–1055. DOI: 10.1021/es902849w 90, 91

Tecon, R., O. Binggeli, et al. (2009). "Double-tagged fluorescent bacterial bioreporter for the study of polycyclic aromatic hydrocarbon diffusion and bioavailability." *Environ. Microbiol.* 11: 2271–2283. DOI: 10.1111/j.1462-2920.2009.01952.x 88, 92

Tecon, R. and J. R. van der Meer (2009). "Effect of two types of biosurfactants on phenanthrene availability to the bacterial bioreporter *Burkholderia sartisoli* strain RP037." *Appl. Microbiol. Biotechnol.* DOI: 10.1007/s00253-009-2216-0 92

Tecon, R., M. Wells, et al. (2006). "A new green fluorescent protein-based bacterial biosensor for analysing phenanthrene fluxes." *Environ. Microbiol.* 8: 697–708. DOI: 10.1111/j.1462-2920.2005.00948.x 87

Tropel, D. and J. R. van der Meer (2004). "Bacterial transcriptional regulators for degradation pathways of aromatic compounds." *Microbiol. Mol. Biol. Rev.* 68(3): 474–500. DOI: 10.1128/MMBR.68.3.474-500.2004 82

Turner, K., S. Xu, et al. (2007). "Hydroxylated polychlorinated biphenyl detection based on a genetically engineered bioluminescent whole-cell sensing system." *Anal. Chem.* 79(15): 5740–5745. DOI: 10.1021/ac0705162 91

van der Meer, J. R., D. Tropel, et al. (2004). "Illuminating the detection chain of bacterial bioreporters." *Environ. Microbiol.* 6: 1005–1020. DOI: 10.1111/j.1462-2920.2004.00655.x 79

Virta, M., J. Lampinen, et al. (1995). "A luminescence-based mercury biosensor." *Anal. Chem.* 67(3): 667–669. DOI: 10.1021/ac00099a027 80

Wells, M., M. Gösch, et al. (2005). "Ultrasensitive reporter protein detection in genetically engineered bacteria." *Anal. Chem.* 77: 2683–2689. DOI: 10.1021/ac048127k 93

Werlen, C., M. C. M. Jaspers, et al. (2004). "Measurement of biologically available naphthalene in gas, and aqueous phases by use of a *Pseudomonas putida* biosensor." *Appl. Environ. Microbiol.* 70: 43–51. DOI: 10.1128/AEM.70.1.43-51.2004 84, 85, 90, 91, 92

Wu, J. and B. P. Rosen (1993). "Metalloregulated expression of the *ars* operon." *J. Biol. Chem.* 268(1): 52–58. 73

CHAPTER 4

Epilogue

4.1 SUMMARY

The development of bacterial bioreporter technology during the past 15-20 years can in hindsight be seen as one of the first applications in synthetic biology. Although fundamentally not as challenging as genome refactoring or metabolic pathway engineering, biosensing with bacterial bioreporters is one of the widely and publicly positively viewed areas of synthetic biology and biotechnology. The idea of assembling parts to simple genetic circuits distilled out of the wide use of a limited number of reporter genes, the expression of which was brought under control of promoter elements and transcriptional regulators from a variety of sources. As the lecture above demonstrates, we still lack many details of even such very simple genetic circuits based on a single transcriptional regulator activating or repressing a single reporter gene from a cognate promoter, which limits the full rational application of the wide variety of existing parts. The reason for this is that most transcriptional regulators for biosensing are significantly distant from well-known regulators such as TetR or LacI, which are still widely popular in fundamental gene network studies but of little relevance when it comes down to detection of toxic compounds. On the other hand, despite gaps in our understanding of the biochemical functioning of many of the used parts, most reporter circuits work quite well in the sense that accurate measurements can be performed and reasonably low detection levels can be achieved for the targeted compound(s). I believe it is this simplicity of construction plus the prospect of developing an assay for detection of a relevant compound which makes biosensing with bacterial bioreporters so attractive.

4.2 FUTURE DIRECTIONS

As I hope to have convinced the reader, biosensing with bacterial and other single cell bioreporters is a mature field of research. Future work will certainly fill in many details in biochemical understanding of used parts, and will further improve the technology of reporter circuit construction in other bacterial species or cell lines. More parts will become available to target a wider variety of chemicals or specific chemicals more precisely, either via mining of bacterial genome and metagenome sequences or via design and direct evolution strategies. The more challenging new direction in bioreporter research will come from a merger with microtechnology. Bioreporter cells are excellently sized to fit microfluidics and other micro-engineered devices, which means that bioassays can be downscaled to fit smaller devices and spare reagents. Microbiosensors can have integrated detectors to record the reporter signals and fluidics could be optimized to provide cells with necessary nutrients to achieve shortest activation time. At the same time, miniaturization could be used to multiple the number of possible targets on a single biosensor device. Multiple reporter strains all based on the same host strain but with different transcription regulator circuits targeting different target compounds can be embedded on the same micro-device. There are published examples of such and other ideas already (van der Meer & Belkin, 2010), and I think it very likely that their number will further increase as more avenues are explored. In combination with the simplicity and low costs of producing bioreporter strains, this will present biosensors as realistic alternatives for a wide range of monitoring applications.

REFERENCES

van der Meer, J. R. and S. Belkin (2010). "Where microbiology meets microengineering: design and applications of reporter bacteria." *Nat. Rev. Microbiol.* 8(7): 511–522. DOI: 10.1038/nrmicro2392

APPENDIX A

Bacterial Bioreporter Designs Targeting Organic Compounds

Compound or compound class	Characteristics	Design specificities
2,4-dichlorophenoxy acetic acid (2,4-D)	Switch	P$_{DII}$ (tfdD$_{II}$) TfdR
	Source genes	tfdD$_{II}$C$_{II}$E$_{II}$F$_{II}$BK (chlorocatechol degradation) – plasmid pJP4
	Source organism	Ralstonia eutropha (Cupriavidus necator) JMP134
	Reporter gene	luxCDABE (V. f.)
	Reporter design	Prom. fusion + regulator:chromosome: random transposon insertion in chromosome (Tn[tfdR -P$_{DII}$::luxCDABE])
	Host strain for construction	Ralstonia eutropha – Homologous - Degradation: +
	Assay	Fresh – Rich medium – 28°C – 100 min induction time
	Measurement	Luminometry
	Detected compounds	2,4-D, 2,4-dichlorophenol, 3-chlorobenzoate
	Concentration range	2 µM - 5 mM (112 µM)
	Application	Aqueous soil extracts and soil slurries from Agent Orange-contaminated soils
	Reference	(Hay, Rice et al. 2000)
Alkanes	Switch	P$_{alkB}$ (alkB) AlkS
	Source genes	alkBFGHJKL (n-alkane degradation) OCT plasmid
	Source organism	P. oleovorans
	r gene	luxAB (V. h.)
	Reporter design	Prom. fusion : plasmid: pJAMA7 (P$_{alkB}$:: luxAB) Regulator:plasmid: pGEc74 (alkST)
	Host strain for construction	E. coli - Heterologous. - Degradation: -
	Assay	Frozen – Rich medium – 30°C – 60 min incubation time
	Measurement	Luminometry
	Detected compounds	pentane, hexane, heptane, octane nonane, decane, 3-methylheptane^y
	Concentration range	25 nM - 6.3 mM
	Application	heating-oil contaminated ground water, oil contaminated sea water
	Reference	(Sticher, Jaspers et al. 1997) (Tecon, Beggah et al. 2010)
Alkanes	Switch	P$_{alkB}$ (alkB) AlkS
	Source genes	alkBFGHJKL (n-alkane degradation) OCT plasmid
	Source organism	P. oleovorans
	Reporter gene	gfp-F64L-S65T(A. v.)
	Reporter design	Prom. fusion + regulator : plasmid: pJAMA30 (AlkS-P$_{alkB}$:: gfp-F64L-S65T)
	Host strain for construction	E. coli - Heterologous. - Degradation: -
	Assay	Fresh – MM + glucose – 30°C – 2.5 h incubation time

Compound or compound class	Characteristics	Design specificities
	Measurement	Fluorometry, Epifluorescence microscopy
	Detected compounds	octane
	Concentration range	10 nM – 1 µM
	Application	Source diffusion tests
	Reference	(Jaspers, Meier et al. 2001)
Benzoates	Switch	P*m* (*xylX*) - XylS
	Source genes	*xylXYZLTEGFJQKIH* (lower pathway for toluene/xylene degradation)
	Source organism	*P. putida* TOL plasmid
	Reporter gene	*luxAB* (*V. harveyi*)
	Reporter design	Prom. fusion + regulator. random transposon insertion in chromosome (Tn[*xylS*-P*m*-::*luxAB*])
	Host strain for construction	*P. putida* - Homologous - Degradation: -
	Assay	Fresh – Rich medium – 30°C – 16 h incubation time
	Measurement	Luminometry
	Detected compounds	benzoate
	Concentration range	5 mM - 100 mM (50 mM)
	Application	Aqueous solutions
	Reference	(de Lorenzo, Fernández et al. 1993)
Benzoates	Switch	P*m* (*xylX*) - XylS
	Source genes	*xylXYZLTEGFJQKIH* (lower pathway for toluene/xylene degradation)
	Source organism	*P. putida* TOL plasmid
	Reporter gene	*gfp*-mut3b (*A. v.*)
	Reporter design	Prom. fusion + regulator. random transposon insertion in chromosome (Tn[*xylS*-P*m*-::*gfp*-mut3b])
	Host strain for construction	*P. putida*- Homologous - Degradation: +
	Assay	Fresh, grown in biofilms – MM – 24 h incubation time
	Measurement	Epiffluorescence microscopy
	Detected compounds	3-methylbenzoate
	Concentration range	5 mM - ND
	Application	Aqueous solutions
	Reference	(Møller, Sternberg et al. 1998)
Chlorobenzoates	Switch	P*fcbA* (*fcbA*) - ND
	Source genes	*fcbABC* (chlorobenzoate degradation)
	Source organism	*Arthrobacter sp*
	Reporter gene	*luxCDABE* (*V. f.*)
	Reporter design	Prom. fusion + regulator. plasmid: pASU (*unknown regulator*-P*fcbA*::*luxCDABE*)
	Host strain for construction	*E. coli* – Heterologous - Degradation: -
	Assay	Fresh – Rich medium – 26°C – 6 h incubation time
	Measurement	Luminometry

Compound or compound class	Characteristics	Design specificities
	Detected compounds	2-/3-/4-chlorobenzoate, 2-/3-/4-bromobenzoate, 4-fluorobenzoate
	Concentration range	0.4 mM – 26 mM
	Application	Industrial wastewater
	Reference	(Rozen, Nejidat et al. 1999)
Hydroxybiphenyls	Switch	P$_{hbpC}$ (*hbpC*) - HbpR
	Source genes	*hbpCA* (*meta*-cleavage pathway of hydroxybiphenyls)
	Source organism	*P. azelaica* HBP1
	Reporter gene	*luxAB* (*V. harveyi*)
	Reporter design	Prom. fusion: chromosome: random transposon insertion in chromosome (Tn[P$_{hbpC}$::*luxAB*]) Regulator: chromosome: host encoded
	Host strain for construction	*P. azelaica* – Homologous - Degradation: +
	Assay	Frozen – MM – 30°C – 75 min incubation time
	Measurement	Luminometry
	Detected compounds	2-hydroxybiphenyl, 2,2'-dihydroxybiphenyl, 2-aminobiphenyl, 2-hydroxydiphenylmethane[i]
	Concentration range	9 µM – 5 mM (0.5 mM)
	Application	Aqueous solutions
	Reference	(Jaspers, Suske et al. 2000)
Hydroxybiphenyls	Switch	P$_{hbpC}$ (*hbpC*) - HbpR
	Source genes	*hbpCA* (*meta*-cleavage pathway of hydroxybiphenyls)
	Source organism	*P. azelaica* HBP1
	Reporter gene	*luxAB* (*V. harveyi*)
	Reporter design	Prom. fusion + regulator: Plasmid: pJAMA8-*hbpR* - P$_{hbpC}$::*luxAB*])
	Host strain for construction	*E. coli* - Heterologous - Degradation: -
	Assay	Fresh – MM + glucose – 30°C – 2-3 h incubation time
	Measurement	Luminometry
	Detected compounds	2-hydroxybiphenyl
	Concentration range	0.4 µM – 8 µM
	Application	Crab urine, Blood plasma
	Reference	(Turner, Xu et al. 2007; Lewis, Beggah et al. 2009)
2-OH and 2-chlorobiphenyl	Switch	P$_{hbpC}$ (*hbpC*) – HbpR-CBP6
	Source genes	*hbpCA* (*meta*-cleavage pathway of hydroxybiphenyls)
	Source organism	*P. azelaica* HBP1 – directed evolution approaches
	Reporter gene	*gfp*-F64L-S65T (*A. v.*)
	Reporter design	Prom. fusion + regulator: Plasmid: pPROBE-*hbpR** - P$_{hbpC}$::*egfp*
	Host strain for construction	*E. coli* - Heterologous - Degradation: -

Compound or compound class	Characteristics	Design specificities
	Assay	Fresh – MM + glucose – 30°C – 2.5-4 h incubation time
	Measurement	Fluorometry, flow cytometry
	Detected compounds	2-hydroxy- and 2-chlorobiphenyl
	Concentration range	1.5 µM – 100 µM
	Application	Aqueous solutions, triclosan, Aroclor
	Reference	(Beggah, Vogne et al. 2008)
Naphthalene	Switch	P_{sal} (*nahG*) - NahR
	Source genes	*nahGTHINLOMKJ* (lower pathway for naphthalene degradation)
	Source organism	*P. fluorescens* plasmid pKA1
	Reporter gene	*luxCDABE* (*V. f.*)
	Reporter design	Prom. fusion + regulator. plasmid: pUTK21, derivative of plasmid pKA1, containing transposon insertion in *nahG* gene (*nahR*-P_{sal}-*nahG'*-Tn[*luxCDABE*])
	Host strain for construction	*P. fluorescens* - Homologous - Degradation: +[j]
	Assay	Fresh (King, DiGrazia et al. 1990; Heitzer, Applegate et al. 1998) Immobilized at the top of a fiber-optic cable (Heitzer, Malachowsky et al. 1994) Immobilised in a packed-bed reactor (Webb, Bienkowski et al. 1997) MM (succinate + yeast extract) (King, DiGrazia et al. 1990) Rich medium (Heitzer, Webb et al. 1992) 25 (King, DiGrazia et al. 1990) 27 (Heitzer, Webb et al. 1992)
	Measurement	Luminometry, fiber optics
	Detected compounds	naphthalene, salicylate
	Concentration range	0.35 µM – 44 µM (Heitzer, Webb et al. 1992)
	Application	aqueous extracts from petroleum contaminated soil (Heitzer, Applegate et al. 1998), aqueous extracts from MPG[k] soil (Heitzer, Malachowsky et al. 1994), aqueous solutions saturated with JP-4 jet fuel (Heitzer, Applegate et al. 1998) (Heitzer, Malachowsky et al. 1994), fuel contaminated soil slurries (Burlage, Palumbo et al. 1994), MPG soil slurries (King, DiGrazia et al. 1990) (Sanseverino, Werner et al. 1993)
	Reference	(King, DiGrazia et al. 1990)
Naphthalene	Switch	P_{nah} (*nahA*) - NahR
	Source genes	*nahABCDEF* (upper pathway for naphthalene degradation)
	Source organism	*P. putida* NAH7 plasmid
	Reporter gene	*luxCDABE* (*V. f.*)
	Reporter design	Prom. fusion: plasmid: pUTK9 (P_{nah}::*luxCDABE*) Regulator: plasmid: NAH7 (*nahR*)
	Host strain for construction	*P. putida* (NAH7) – Homologous - Degradation: +
	Assay	Fresh – MM or Rich medium – 25°C. Incubation times: 120 - 240 min (MM) 16 - 20h (rich medium)
	Measurement	Luminometry
	Detected compounds	naphthalene, salicylate
	Concentration range	0.8 mM[l]
	Application	Aqueous solutions
	Reference	(Burlage, Sayler et al. 1990)

Compound or compound class	Characteristics	Design specificities
Naphthalene	Switch	P*sal* (*nahG*) NahR
	Source genes	*nahGTHINLOMKJ* (lower pathway for naphthalene degradation)
	Source organism	*P. putida* NAH7 plasmid
	Reporter gene	*luxAB* (*V. h*)
	Reporter design	Prom. fusion: mini-transposon Tn[*nahR'-Psal::luxAB*} on chromosome Regulator: Plasmid NAH7 (*nahR*)
	Host strain for construction	*P. putida* (NAH7) – Homologous - Degradation: +
	Assay	Fresh (chemostat grown cells), aqueous assay in MM – 30°C, 2-3 h incubation time
	Measurement	Luminometry
	Detected compounds	Naphthalene, 2-methyl-, 2,3-dimethylnaphthalene, salicylate
	Concentration range	0.5 µM – 5 µM (aqueous phase assay) 0.05 – 0.5 µM (gas phase assay)
	Application	Oil contaminated sites and wastewaters
	Reference	(Werlen, Jaspers et al. 2004)
Phenanthrene	Switch	P*phnS* (*phnS*) - PhnR
	Source genes	*phnSFECDAcAbB* (upper pathway of phenanthrene degradation)
	Source organism	*Burkholderia sartisoli* RP007
	Reporter gene	*gfp*-F64L-S65T (*A. v.*)
	Reporter design	Prom. fusion: *phnS'-egfp* on pPROBE Regulator: *phnR* (chromosome)
	Host strain for construction	*B. sartisoli* RP037 – Homologous - degradation: +
	Assay	Fresh, MM + acetate – 30°C; aqueous assays or diffusion assays, 1-3d incubation time
	Measurement	Single cell EFM
	Detected compounds	Phenanthrene, naphthalene
	Concentration range	Flux range: 2 – 70 fg/cell
	Application	Contaminated soils or diffusion assays
	Reference	(Tecon, Wells et al. 2006) (Tecon, Binggeli et al. 2009)
Phenanthrene	Switch	P*phnS* (*phnS*) - PhnR
	Source genes	*phnSFECDAcAbB* (upper pathway of phenanthrene degradation)
	Source organism	*Burkholderia sartisoli* RP007
	Reporter gene	*luxAB* (*V.h.*)
	Reporter design	Prom. fusion: *phnS'-egfp* on pPROBE-*lux* Regulator:*phnR* (chromosome)
	Host strain for construction	*B. sartisoli* RP037 – Homologous - degradation: +
	Assay	Fresh, MM – 30°C; aqueous assays, 3 h incubation time
	Measurement	Luminometry
	Detected compounds	naphthalene
	Concentration range	0.17 – 10 µM
	Application	Oil contaminated seawater
	Reference	(Tecon, Beggah et al. 2010)

Compound or compound class	Characteristics	Design specificities
Phenols	Switch	Po (*dmpK*) - DmpR
	Source genes	*dmpKLMNOPQBCDEFGHI* (phenol degradation)
	Source organism	*Pseudomonas* sp. Plasmid pVI150
	Reporter gene	*luxAB* - (*V. h.*)
	Reporter design	Prom. fusion: plasmid: pVI360 (Po::luxAB) Regulator: chromosome: host encoded
	Host strain for construction	*P. putida*::*dmpR* – Homologous - Degradation: -
	Assay	Fresh, MM – 30°C, 90 min incubation time
	Measurement	Luminometry
	Detected compounds	phenol, 2-/3-/4-methylphenol, 3,4-dimethylphenol[†]
	Concentration range	3.2 µM - 2 mM
	Application	Aqueous solutions
	Reference	(Shingler and Moore 1994)
Polychlorinated biphenyls	Switch	P$_{bphA1}$ (*bphA1*)[k] ND
	Source genes	*bphA1A2(orf3)bphA3A4BC* (meta cleavage pathway of (chloro)biphenyls)
	Source organism	*Ralstonia eutropha*
	Reporter gene	*luxCDABE* (*V. f.*)
	Reporter design	Prom. fusion: plasmid: pUTK60 (*orf0*-P$_{bphA1}$-*bphA1*::*luxCDABE*) Regulator: chromosome: host encoded
	Host strain for construction	*Ralstonia eutropha* – Homologous - Degradation: +
	Assay	Fresh, MM + pyruvate – 26°C, 120 min incubation time
	Measurement	Luminometry
	Detected compounds	biphenyl, 2-/3-/4-chlorobiphenyl
	Concentration range	0.79 µM – 0.11 mM
	Application	Aqueous suspensions
	Reference	{Layton, 1998 #553
Styrene	Switch	P$_{styA}$ StyS/R
	Source genes	*styABCDE* (styrene monooxygenation)
	Source organism	*Pseudomonas* sp. Strain Y2
	Reporter gene	*luxCDABE* (*V. f.*)
	Reporter design	Prom. Translational fusion + regulator: chromosome: random transposon insertion in chromosome (Tn[P$_{sty}$::*lacZ*])
	Host strain for construction	*Pseudomonas* sp. Strain Y2 PAL1 – Homologous - Degradation: +
	Assay	Fresh – M9 glycerol medium– 28°C – 60 min induction time
	Measurement	Spectrophotometry
	Detected compounds	Styrene, toluene, epoxystyrene, phenylacetaldehyde and 2-phenylethanol
	Concentration range	1 µM – 0.1 mM (100 µM)
	Application	Aqueous suspensions
	Reference	(Alonso, Navarro-Llorens et al. 2003)

Compound or compound class	Characteristics	Design specificities
Toluene	Switch	P_{tod} (*todX*) TodST
	Source genes	*todXFC1C2BADEGHI* (degradation of toluene)
	Source organism	*P. putida* F1
	Reporter gene	*luxCDABE* (*V. f.*)
	Reporter design	Prom. fusion: plasmid: pUTK30 (P_{tod}-*todXF'::luxCDABE*) Regulator: chromosome: host encoded
	Host strain for construction	*P. putida* F1- Homologous - Degradation: +
	Assay	Fresh; Immobilised in a packed-bed reactor, Rich medium – 27°C, 90 min incubation time
	Measurement	Luminometry
	Detected compounds	toluene
	Concentration range	1 µM – 0.5 mM
	Application	aqueous solutions saturated with JP-4 jet fuel
	Reference	(Applegate, Kelly et al. 1997)
Toluene	Switch	P_{tod} (*todX*) TodST
	Source genes	*todXFC1C2BADEGHI* (degradation of toluene)
	Source organism	*P. putida* F1
	Reporter gene	*luxCDABE* (*V. f.*)
	Reporter design	Prom. fusion: chromosome: random transposon insertion in chromosome (Tn[P_{tod}-*todXF'::luxCDABE*)]) Regulator: chromosome: host encoded
	Host strain for construction	*P. putida* F1- Homologous - Degradation: +
	Assay	Fresh, MM – 28°C, 120 min incubation time
	Measurement	Luminometry
	Detected compounds	toluene, benzene, 2-/3-xylene, phenol, trichloroethylene (Shingleton, Applegate et al. 1998)
	Concentration range	0.3 µM – 0.5 mM
	Application	aqueous solutions saturated with JP-4 jet fuel
	Reference	(Applegate, Kehrmeyer et al. 1998)
Toluene	Switch	P_{tbuA1} (*tbuA1*) - TbuT
	Source genes	*tbuA1UBVA2CT* (upper pathway for toluene/benzene degradation)
	Source organism	*R. picketti* PKO1
	Reporter gene	*gfp* (red shifted variant) (*A. v.*)
	Reporter design	Prom. fusion: P_{tbuA1}::*gfp* (location not described) Regulator: *tbuT* driven expression from constitutive promoter (location not described)
	Host strain for construction	*P. fluorescens* A506 – Heterologous – Degradation: ND
	Assay	Fresh, ND, ND, 3 h incubation time
	Measurement	Epifluorescence microscopy, fluorometry
	Detected compounds	Toluene, benzene, trichloroethylene
	Concentration range	20 nM - 2 µM
	Application	rhizosphere
	Reference	(Halverson 1999)

Compound or compound class	Characteristics	Design specificities
Toluene	Switch	P_{tbuA1} (*tbuA1*) - TbuT
	Source genes	*tbuA1UBVA2CT* (upper pathway for toluene/benzene degradation)
	Source organism	*R. picketti* PKO1
	Reporter gene	*luxAB (V. h.)*
	Reporter design	Prom. fusion + regulator: pPROBE-*tbuT*-P_{tbuA1}::*luxAB*
	Host strain for construction	*E. coli* – Heterologous - Degradation: -
	Assay	Frozen, MM + glucose, 30°C, 2 h incubation time
	Measurement	Luminometry
	Detected compounds	Toluene
	Concentration range	0.24 – 10 µM
	Application	Oil contaminated marine waters
	Reference	(Tecon, Beggah et al. 2010)
Xylenes	Switch	P*u* (*xylU*) - XylR
	Source genes	*xylUWCMABN* (upper pathway for toluene/xylene degradation)
	Source organism	*P. putida* mt-2 TOL plasmid
	Reporter gene	*luxCDABE (V. f.)*
	Reporter design	Prom. fusion + regulator: plasmid: undescribed plasmid (P*u*:::*luxCDABE*) Regulator: plasmid: TOL plasmid
	Host strain for construction	*P. putida* (TOL) – Homologous - Degradation: +
	Assay	Fresh, MM, 27°C, 120 min incubation time
	Measurement	Luminometry
	Detected compounds	toluene, 2-/3-/4-xylene
	Concentration range	15 µM^n(ND)
	Application	Fuel contaminated soil slurries
	Reference	(Burlage, Palumbo et al. 1994)
Xylenes	Switch	P*u* (*xylU*) - XylR
	Source genes	*xylUWCMABN* (upper pathway for toluene/xylene degradation)
	Source organism	*P. putida* mt-2 TOL plasmid
	Reporter gene	*gfp* mut3 (*A. v.*)
	Reporter design	Prom. fusion + regulator: chromosome: random transposon insertion in chromosome (Tn[*xylR*-P*u*-::*gfp*-mut3b])
	Host strain for construction	*P. putida* – homologous– Degradation: +
	Assay	Fresh, grown in biofilms , MM, ND, 24 h incubation time
	Measurement	Epifluorescence microscopy
	Detected compounds	benzyl alcohol
	Concentration range	0.26 mM (ND)
	Application	Aqueous solution
	Reference	(Møller, Sternberg et al. 1998)

Compound or compound class	Characteristics	Design specificities
Xylenes	Switch	Pu (*xylU*) - XylR
	Source genes	*xylUWCMABN* (upper pathway for toluene/xylene degradation)
	Source organism	*P. putida* mt-2 TOL plasmid
	Reporter gene	*lucFF* (P. p.)
	Reporter design	Prom. fusion + regulator plasmid: pGLTUR(*xylR*-Pu:::*lucFF*)
	Host strain for construction	*E. coli* – Heterologous - Degradation: -
	Assay	Fresh, Rich medium, 37°C, 30 min incubation time
	Measurement	Luminometry
	Detected compounds	toluene, benzene, 2-/3-/4-xylene, 2-/3-/4-chlorotoluene, 2-/3-/4-methylbenzylalcohol, 2-/3-/4-nitrotoluene
	Concentration range	10 µM – 1 mM (ND)
	Application	BETX^p contaminated deep aquifer water; ethyl alcohol extracts of BETX contaminated soil
	Reference	(Willardson, Wilkins et al. 1998)
Xylenes	Switch	Pu (*xylU*) - XylR
	Source genes	*xylUWCMABN* (upper pathway for toluene/xylene degradation)
	Source organism	*P. putida* mt-2 TOL plasmid
	Reporter gene	*luxCDABE* (V.f.)
	Reporter design	Prom. fusion + regulator plasmid: pTOLLUX (*xylR*-Pu :::*luxCDABE*)
	Host strain for construction	*E. coli* – Heterologous - Degradation: -
	Assay	Fresh, Rich medium, 37°C, 2 h incubation time
	Measurement	Luminometry
	Detected compounds	toluene, 2-/3-/4-xylene, benzylaldehyde, naphthalene
	Concentration range	7.5 µM – 1 mM
	Application	Contaminated groundwaters or soil-water extracts
	Reference	(Li, Li et al. 2008)
Xylenes	Switch	Ps (*xylS*) - XylR
	Source genes	*xylS* (regulation of the *xylXYZLTEGFJQKIH* genes, encoding the lower pathway for toluene/xylene degradation)
	Source organism	*P. putida* mt-2 TOL plasmid
	Reporter gene	*lucFF* (P. p.)
	Reporter design	Prom. fusion + regulator: plasmid: pTSN316 (*xylR*-Ps::*lucFF*)
	Host strain for construction	*E. coli* – Heterologous - Degradation: -
	Assay	Fresh (Kobatake, Niimi et al. 1995) Immobilized at the top of a fiber-optic cable (Ikariyama, Nishiguchi et al. 1997) , Rich medium, 37°C, 120 min incubation
	Measurement	Luminometry
	Detected compounds	benzene, toluene, ethylbenzene, 2-/3-/4-xylene, 3-ethyltoluene, 2-/3-/4-chlorotoluene

Compound or compound class	Characteristics	Design specificities
	Concentration range	5 µM - 1 mM (0.5 mM)
	Application	Contaminated groundwaters or soil-water extracts
	Reference	(Kobatake, Niimi et al. 1995)
Various hydrophobic compounds	Switch	P$_{ibpA}$ (*ibpA*) - IbpR
	Source genes	*ibpACDABE* (alkylbenzene degradation)
	Source organism	*P. putida*
	Reporter gene	*luxCDABE (V.f.)*
	Reporter design	Prom. fusion + regulator: plasmid: pOS25 (*ibpR*-P$_{ibpA}$::*luxCDABE*)
	Host strain for construction	*E. coli* – Heterologous - Degradation: -
	Assay	Fresh, MM + pyruvate, 22°C, 250 min incubation time
	Measurement	Luminometry
	Detected compounds	<u>isopropylbenzene</u>, benzene, toluene, ethylbenzene, *n*-butylbenzene, naphthalene, trichloroethylene[q]
	Concentration range	1 µM - 1mM (0.1 mM)
	Application	gasoline, jet fuel JP-4 and JP-5, diesel fuel, creosote, ethanol extract of hydrocarbon contaminated lake sediment
	Reference	(Selifonova and Eaton 1996)
Lysine	Switch	Lysine auxotroph
	Source genes	-
	Source organism	*E. coli*
	Reporter gene	*Gfpmut3*
	Reporter design	Prom. fusion: constitutively expressed *gfp* from mini-Tn*5* inserted into *E. coli lysA*⁻
	Host strain for construction	*E. coli* - Homologous, degradation: +
	Assay	Fresh, M9, 37°C, 6 h incubation time
	Measurement	Fluorometry
	Detected compounds	<u>Lysine</u>
	Concentration range	0.5 – 2 mg / ml
	Application	Food stuffs
	Reference	(Chalova, Woodward et al. 2006)
Tetracycline	Switch	P*tet* - TetR
	Source genes	*tetA* (tetracycline resistance)
	Source organism	*E. coli*
	Reporter gene	*Gfpmut3*
	Reporter design	Prom. fusion: plasmid pTGFP2 (P*tet*-*gfp*)
	Host strain for construction	*E. coli* – Homologous - Degradation: -
	Assay	Fresh (intestinal colonization), 1-7 d incubation time – faeces analysis
	Measurement	Flow cytometry
	Detected compounds	<u>tetracycline</u>
	Concentration range	0.02 – 0.4 mg/l

Compound or compound class	Characteristics	Design specificities
	Application	Rat intestine exposure
	Reference	(Bahl, Hansen et al. 2004)
N-Acyl homoserine lactones	Switch	P*luxI* - LuxR
	Source genes	*luxI (*Luciferase quorum sensing system)
	Source organism	*Photorabdus luminescens*
	Reporter gene	*Gfp*mut3
	Reporter design	Prom- fusion + regulator: plasmid pAHL-GFP (*luxR-P_{luxI}-gfp*mut3)
	Host strain for construction	*E. coli* – Heterologous - Degradation: -
	Assay	Fresh, Rich medium, 37°C (aqueous), 25°C (soil microcosms), 20 h incubation time
	Measurement	Flow cytometry
	Detected compounds	N-octanoyl homoserine lactone
	Concentration range	0.1 – 10 µM (aqueous) (10 µM), 0.5 nM / g soil
	Application	Solutions, soils, tissue
	Reference	(Burmølle, Hansen et al. 2003)
Various organic compounds	Switch	P*sepA* - SepR
	Source genes	*sepABC* (solvent efflux system)
	Source organism	*P. putida* F1
	Reporter gene	*luxCDABE (P. l.)*
	Reporter design	Prom. fusion: Mini-Tn5 [P*sepA-luxCDABE*) Regulator: Chromosome
	Host strain for construction	*P. putida F1* - homologous, degradation: +
	Assay	Fresh, M9 plus glucose, 30°C, 2 h incubation time
	Measurement	Luminometry
	Detected compounds	Benzene, isopropylbenzene, 4-ethyl-, 4-chlorotoluene, naphthalene, styrene (and others)
	Concentration range	0.5 – 20 mM (aqueous)
	Application	Jet-fuel, oil contaminations
	Reference	(Phoenix, Keane et al. 2003)
2,4-dinitrotoluene	Switch	P*u* - XylR5
	Source genes	*xylU* (upper pathway for toluene degradation) - directed evolution
	Source organism	*P. putida* mt-2 (directed evolution on XylR)
	Reporter gene	*luxCDABE (P. l.)*
	Reporter design	Prom. fusion + regulator: Mini-Tn5 [*xylR5-Pu-luxCDABE*)
	Host strain for construction	*P. putida* KT2400, homologous, degradation: -
	Assay	Fresh, rich medium, 30°C, vapour, 20 h assay time
	Measurement	CCD imaging
	Detected compounds	2,4-dinitrotoluene, salicylate
	Concentration range	100 mg
	Application	Mine detection
	Reference	(de las Heras, Carreno et al. 2008)

Compound or compound class	Characteristics	Design specificities
	Host strain for construction	*P. putida* KT2400, homologous, degradation: -
	Assay	Fresh, rich medium, 30°C, vapour, 20 h assay time
	Measurement	CCD imaging
	Detected compounds	2,4-dinitrotoluene, salicylate
	Concentration range	100 mg
	Application	Mine detection
	Reference	(de las Heras, Carreno et al. 2008)

REFERENCES

Alonso, S., J. M. Navarro-Llorens, et al. (2003). "Construction of a bacterial biosensor for styrene." *J. Biotechnol.* **102**(3): 301–306. DOI: 10.1016/S0168-1656(03)00048-8

Applegate, B., C. Kelly, et al. (1997). "*Pseudomonas putida* B2: a *tod-lux* bioluminescent reporter for toluene and trichloroethylene co-metabolism." *J. Ind. Microbiol.* **18**: 4–9.

Applegate, B. M., S. R. Kehrmeyer, et al. (1998). "A chromosomally based *tod-luxCDABE* whole-cell reporter for benzene, toluene, ethylbenzene, and xylene (BTEX) sensing." *Appl. Environ. Microbiol.* **64**(7): 2730–2735.

Bahl, M. I., L. H. Hansen, et al. (2004). "In vivo detection and quantification of tetracycline by use of a whole-cell biosensor in the rat intestine." *Antimicrob. Agents Chemother.* **48**(4): 1112–1117. DOI: 10.1128/AAC.48.4.1112-1117.2004

Beggah, S., C. Vogne, et al. (2008). "Mutant transcription activator isolation via green fluorescent protein based flow cytometry and cell sorting." *Microb. Biotechnol.* **1**: 68–78.

Burlage, R. S., A. V. Palumbo, et al. (1994). "Bioluminescent reporter bacteria detect contaminants in soil samples." *Appl. Biochem. Biotechnol.* **46**(0): 731–740. DOI: 10.1007/BF02941845

Burlage, R. S., G. S. Sayler, et al. (1990). "Monitoring of naphthalene catabolism by bioluminescence with *nah-lux* transcriptional fusions." *J. Bacteriol.* **172**(9): 4749–4757.

Burmølle, M., L. H. Hansen, et al. (2003). "Presence of N-acyl homoserine lactones in soil detected by a whole-cell biosensor and flow cytometry." *Microb. Ecol.* **45**: 226–236. DOI: 10.1007/s00248-002-2028-6

Chalova, V., C. L. Woodward, et al. (2006). "Application of an *Escherichia coli* green fluorescent protein-based lysine biosensor under nonsterile conditions and autofluorescence background." *Lett. Appl. Microbiol.* **42**(3): 265–270. DOI: 10.1111/j.1472-765X.2005.01834.x de las Heras, A., C. A. Carreno, et al. (2008). "Stable implantation of orthogonal sensor circuits in Gram-negative bacteria for environmental release." *Environ. Microbiol.* **10**(12): 3305–3316. DOI: 10.1111/j.1462-2920.2008.01722.x

de Lorenzo, V., S. Fernández, et al. (1993). "Engineering of alkyl- and haloaromatic-responsive gene expression with mini-transposons containing regulated promoters of biodegradative pathways of *Pseudomonas*." *Gene* **130**(1): 41–46. DOI: 10.1016/0378-1119(93)90344-3

Halverson, L. J. (1999). "Development of a GFP-based whole-cell biosensor for pollutant detection". ASM conference on *Pseudomonas* '99: biotechnology and pathogenesis, Maui, Hawaii, American Society for Microbiology.

Hay, A. G., J. F. Rice, et al. (2000). "A bioluminescent whole-cell reporter for detection of 2,4-dichlorophenoxyacetic acid and 2,4-dichlorophenol in soil." *Appl. Environ. Microbiol.* **66**: 4589–4594. DOI: 10.1128/AEM.66.10.4589-4594.2000

Heitzer, A., B. Applegate, et al. (1998). "Physiological considerations of environmental applications of *lux* reporter fusions." *J. Microbiol. Methods* **33**: 45–57. DOI: 10.1016/S0167-7012(98)00043-8

Heitzer, A., K. Malachowsky, et al. (1994). "Optical biosensor for environmental on-line monitoring of naphthalene and salicylate bioavailability with an immobilized bioluminescent catabolic reporter bacterium." *Appl. Environ. Microbiol.* **60**(5): 1487–1494.

Heitzer, A., O. F. Webb, et al. (1992). "Specific and quantitative assessment of naphthalene and salicylate bioavailability by using a bioluminescent catabolic reporter bacterium." *Appl. Environ. Microbiol.* **58**(6): 1839–1846.

Ikariyama, Y., S. Nishiguchi, et al. (1997). "Fiber-optic-based biomonitoring of benzene derivatives by recombinant *E. coli* bearing luciferase gene-fused TOL-plasmid immobilized on the fiber-optic end." *Anal. Chem.* **69**(13): 2600–2605. DOI: 10.1021/ac961311o

Jaspers, M. C., W. A. Suske, et al. (2000). "HbpR, a new member of the XylR/DmpR subclass within the NtrC family of bacterial transcriptional activators, regulates expression of 2-hydroxybiphenyl metabolism in *Pseudomonas azelaica* HBP1." *J. Bacteriol.* **182**(2): 405–417.

Jaspers, M. C. M., C. Meier, et al. (2001). "Measuring mass transfer processes of octane with the help of an *alkS-alkB::gfp*-tagged *Escherichia coli*." *Environ. Microbiol.* **3**(8): 512–524. DOI: 10.1046/j.1462-2920.2001.00218.x

King, J. M. H., P. M. DiGrazia, et al. (1990). "Rapid, sensitive bioluminescent reporter technology for naphthalene exposure and biodegradation." *Science* **249**: 778–781. DOI: 10.1126/science.249.4970.778

Kobatake, E., T. Niimi, et al. (1995). "Biosensing of benzene derivatives in the environment by luminescent *Escherichia coli.*" *Biosens. Bioelectr.* **10**(6–7): 601-605. DOI: 10.1016/0956-5663(95)96936-S

Lewis, C., S. Beggah, et al. (2009). "Novel use of a whole cell *E. coli* bioreporter as a urinary exposure biomarker." *Environ. Sci. Technol.* **43**(2): 423–428. DOI: 10.1021/es801325u

Li, Y. F., F. Y. Li, et al. (2008). "Construction and comparison of fluorescence and bioluminescence bacterial biosensors for the detection of bioavailable toluene and related compounds." *Environ. Pollut.* **152**(1): 123–129. DOI: 10.1016/j.envpol.2007.05.002

Møller, S., C. Sternberg, et al. (1998). "*In situ* gene expression in mixed-culture biofilms: evidence of metabolic interactions between community members." *Appl. Environ. Microbiol.* **64**(2): 721–732.

Phoenix, P., A. Keane, et al. (2003). "Characterization of a new solvent-responsive gene locus in *Pseudomonas putida* F1 and its functionalization as a versatile biosensor." *Environ. Microbiol.* **5**: 1309–1327. DOI: 10.1111/j.1462-2920.2003.00426.x

Rozen, Y., A. Nejidat, et al. (1999). "Specific detection of *p*-chlorobenzoic acid by *Escherichia coli* bearing a plasmid-borne *fcbA'::lux* fusion." *Chemosphere* **38**(3): 633–641. DOI: 10.1016/S0045-6535(98)00210-0

Sanseverino, J., C. Werner, et al. (1993). "Molecular diagnostics of polycyclic aromatic hydrocarbon biodegradation in manufactured gas plant soils." *Biodegradation* **4**: 303–321. DOI: 10.1007/BF00695976

Selifonova, O. V. and R. W. Eaton (1996). "Use of an *ipb-lux* fusion to study regulation of the isopropylbenzene catabolism operon of *Pseudomonas putida* RE204 and to detect hydrophobic pollutants in the environment." *Appl. Environ. Microbiol.* **62**(3): 778–783.

Shingler, V. and T. Moore (1994). "Sensing of aromatic compounds by the DmpR transcriptional activator of phenol-catabolizing *Pseudomonas* sp. strain CF600." *J. Bacteriol.* **176**(6): 1555–1560.

Shingleton, J. T., B. M. Applegate, et al. (1998). "Induction of the *tod* operon by trichloroethylene in *Pseudomonas putida* TVA8." *Appl. Environ. Microbiol.* **64**(12): 5049–5052.

Sticher, P., M. Jaspers, et al. (1997). "Development and characterization of a whole cell bioluminescent sensor for bioavailable middle-chain alkanes in contaminated groundwater samples." *Appl. Environ. Microbiol.* **63**(10): 4053–4060.

Tecon, R., S. Beggah, et al. (2010). "Development of a multistrain bacterial bioreporter platform for the monitoring of hydrocarbon contaminants in marine environments." *Environ. Sci. Technol.* **144**: 1049–1055. DOI: 10.1021/es902849w

Tecon, R., O. Binggeli, et al. (2009). "Double-tagged fluorescent bacterial bioreporter for the study of polycyclic aromatic hydrocarbon diffusion and bioavailability." *Environ. Microbiol.* **11**: 2271–2283. DOI: 10.1111/j.1462-2920.2009.01952.x

Tecon, R., M. Wells, et al. (2006). "A new green fluorescent protein-based bacterial biosensor for analysing phenanthrene fluxes." *Environ. Microbiol.* **8**: 697–708. DOI: 10.1111/j.1462-2920.2005.00948.x

Turner, K., S. Xu, et al. (2007). "Hydroxylated polychlorinated biphenyl detection based on a genetically engineered bioluminescent whole-cell sensing system." *Anal. Chem.* **79**(15): 5740–5745. DOI: 10.1021/ac0705162

Webb, O. F., P. R. Bienkowski, et al. (1997). "Kinetics and response of a *Pseudomonas fluorescens* HK44 biosensor." *Biotechnol. Bioeng.* **54**(5): 491–502. DOI: 10.1002/(SICI)1097-0290(19970605)54:5%3C491::AID-BIT8%3E3.0.CO;2-9

Werlen, C., M. C. M. Jaspers, et al. (2004). "Measurement of biologically available naphthalene in gas, and aqueous phases by use of a *Pseudomonas putida* biosensor." *Appl. Environ. Microbiol.* **70**: 43–51. DOI: 10.1128/AEM.70.1.43-51.2004

Willardson, B. M., J. F. Wilkins, et al. (1998). "Development and testing of a bacterial biosensor for toluene-based environmental contaminants." *Appl. Environ. Microbiol.* **64**(3): 1006–1012.

APPENDIX B

Bacterial Bioreporter Designs Targeting (Heavy) Metals and Metalloids

Compound	Nature	Specificities
Arsenic	Switch	P_{ars} (*arsR*) - ArsR
	Source genes	*arsRDABC* (arsenate/arsenite resistance)
	Source organism	*E. coli* plasmid R773
	Reporter gene	*lucFF* (*P. p.*)
	Reporter design	Prom. fusion + regulator: plasmid pTOO31 (*arsR*-P_{ars}::*lucFF*)
	Host strain for construction	*E. coli* – Homologous - Resistance: ±[m]
	Assay	Lyophilized, MM + casein, 30°C, 90 min incubation time
	Measurement	Luminometry
	Detected compounds	$\underline{AsO_2^-}$, AsO_4^{3-}, SbO_2^-, Cd^{2+}
	Concentration range	33 nM - 1 mM (10 mM)
	Application	Aqueous solutions
	Reference	(Tauriainen, Virta et al. 1999)
Arsenic	Switch	P_{ars} (*arsR*) - ArsR
	Source genes	*arsRBC* (arsenate/arsenite resistance)
	Source organism	*E. coli* chromosome
	Reporter gene	*luxAB* (*V. h.*)
	Reporter design	Prom. fusion + regulator: chromosome: transposon insertion in *arsB* gene (*arsR*-P_{ars}-*arsB'*-Tn[*luxAB*])
	Host strain for construction	*E. coli* - Homologous - Resistance: +
	Assay	Fresh, rich medium, 37°C, 60 min induction time
	Measurement	Luminometry
	Detected compounds	$\underline{AsO_2^-}$
	Concentration range	80 nM - 2 mM (0.8 mM)
	Application	chromated copper arsenate (wood preservative) (Cai and DuBow 1997)
	Reference	(Cai and DuBow 1996)
Arsenic	Switch	P_{ars} (*arsR*) - ArsR
	Source genes	*arsRBC* (arsenate/arsenite resistance)
	Source organism	*Staphylococcus aureus* plasmid pI258
	Reporter gene	*luxAB* (*V. h.*)
	Reporter design	Prom. fusion + regulator: plasmid: pC200 (P_{ars}-*arsRB'*::*luxAB*)
	Host strain for construction	*S. aureus* – Homologous - Resistance: +
	Assay	Fresh, rich medium, 37°C, 120 min assay time
	Measurement	Luminometry
	Detected compounds	AsO_2^-

Compound	Nature	Specificities
	Concentration range	1 mM - 20 mM (5 mM)
	Application	Aqueous solution
	Reference	(Corbisier, Ji et al. 1993)
Arsenic	Switch	P$_{ars}$ (*arsR*) - ArsR
	Source genes	*arsRBC* (arsenate/arsenite resistance)
	Source organism	*Staphylococcus aureus* plasmid pl258
	Reporter gene	*lucFF* (*P. p.*)
	Reporter design	Prom. fusion + regulator: plasmid: pTOO21 (P$_{ars}$-*arsR*::*lucFF*)
	Host strain for construction	*S. aureus* – Homologous - Resistance: +
	Assay	Fresh, Lyophilized, MM + casein, 30°C, 120 min assay time
	Measurement	Luminometry
	Detected compounds	AsO$_2^-$, SbO$_2^-$, AsO$_4^{3-}$, Cd^{2+}
	Concentration range	0.1 µM - 10 mM (3.3 mM)
	Application	Aqueous solutions
	Reference	(Tauriainen, Karp et al. 1997)
Arsenic	Switch	P$_{ars}$ (*arsR*) - ArsR
	Source genes	*arsRDABC* (arsenate/arsenite resistance)
	Source organism	*E. coli* plasmid R773
	Reporter gene	*luxAB* (*V. h.*)
	Reporter design	Prom. fusion + regulator: plasmid: pRLUX (P$_{ars}$-*arsRD'*::*luxAB*)
	Host strain for construction	*E. coli* – Homologous - Resistance: ±^{m}
	Assay	Fresh, rich medium, 37°C, 180 min assay time
	Measurement	Luminometry, fiber optics
	Detected compounds	AsO$_2^-$, SbO$_2^-$
	Concentration range	1 aM - 1 mM (10 nM)
	Application	Aqueous solutions
	Reference	(Ramanathan, Shi et al. 1997)
Arsenic	Switch	P$_{ars}$ (*arsR*) - ArsR
	Source genes	*arsRDABC* (arsenate/arsenite resistance)
	Source organism	*E. coli* plasmid R773
	Reporter gene	*luxAB* (*V. h.*) or *egfp* or *lacZ* or *ccp*
	Reporter design	Prom. fusion + regulator: with or without extra ArsR binding site on plasmid
	Host strain for construction	*E. coli* – Homologous - Resistance: +
	Assay	Fresh or from frozen stocks, MM medium with Glu, 30-37°C, 2-4 h assay time
	Measurement	Luminometry, fluorimetry, histology
	Detected compounds	AsO$_2^-$, AsO$_4^-$, SbO$_2^-$
	Concentration range	10 nM – 4 µM linear
	Application	Aqueous solutions, groundwater, food stuffs
	Reference	(Stocker, Balluch et al. 2003; Baumann and van der Meer 2007; Wackwitz, Harms et al. 2008)

Compound	Nature	Specificities
Arsenic	Switch	P_{ars} (*arsR*) - ArsR
	Source genes	*arsRDABC* (arsenate/arsenite resistance)
	Source organism	*E. coli* plasmid R773
	Reporter gene	*luxCDABE* (*V. h.*) or *lucFF* or *gfp*
	Reporter design	Prom. fusion + regulator: plasmids parsRlucFFGFP, parsRGFP, larsRluxCDABE
	Host strain for construction	*E. coli* MC1061 – Heterologous – Resistance: +/-
	Assay	Fresh, M9 plus casein, 37°C, 0.5- 2 h
	Measurement	Luminometry (or fluorometry)
	Detected compounds	As(III)
	Concentration range	30 nM – 1 µM (*luc* or *lux*)
	Application	Aqueous solutions
	Reference	(Hakkila, Maksimow et al. 2002)
Cadmium	Switch	P_{cadC} (*cadC*) - CadC (repressor)
	Source genes	*cadCA* (cadmium resistance)
	Source organism	*S. aureus* plasmid pI258
	Reporter gene	*lucFF* (*P. p.*)
	Reporter design	Prom. fusion + regulator: plasmid: pTOO24 (P_{cadC}-*cadC*::*lucFF*)
	Host strain for construction	*B. subtilis* – Heterologous – Resistance: -
	Assay	Lyophilized, MM (casein), 30°C, 120 min assay time
	Measurement	Luminometry
	Detected compounds	Cd^{2+}, Pb^{2+}, SbO_2^-, Sn^{2+}
	Concentration range	10 nM – 10 µM
	Application	
	Reference	(Tauriainen, Karp et al. 1998)
Cadmium	Switch	P_{cadC} (*cadC*) - CadC (repressor)
	Source genes	*cadCA* (cadmium resistance)
	Source organism	*S. aureus* plasmid pI258
	Reporter gene	*egfp* of pPROBE-NT
	Reporter design	Prom. fusion + regulator: plasmid: pVLCD1 (P_{cadC}-*cadCA'*::*gfp*)
	Host strain for construction	*E. coli* - Heterologous - Resistance: -
	Assay	Fresh, rich medium, 37°C, 120 min assay time
	Measurement	Fluorometry
	Detected compounds	Cd^{2+}, Sb^{3+}, $Pb^{2+,}$ Zn^{21}
	Concentration range	10 nM – 10 µM
	Application	Soil water extracts
	Reference	(Liao, Chien et al. 2006)
Cadmium	Switch	P_{cadC} (*cadC*) - CadC (repressor)
	Source genes	*cadCA* (cadmium resistance)
	Source organism	*S. aureus* plasmid pI258
	Reporter gene	*luxAB* (*V. h.*)

Compound	Nature	Specificities
	Reporter design	Prom. fusion + regulator: plasmid: pC300 (P_{cadC}-*cadCA*::*luxAB*)
	Host strain for construction	*S. aureus* – Homologous - Resistance: +
	Assay	Fresh, rich medium, 37°C, 120 min assay time
	Measurement	Luminometry
	Detected compounds	Cd^{2+}, Bi^{3+}, Pb^{2+}
	Concentration range	0.5 mM - 0.1 mM (20 mM)
	Application	Aqueous solutions or suspensions
	Reference	(Corbisier, Ji et al. 1993)
Cadmium	Switch	P_{zntA} - ZntR
	Source genes	*zntRA* (cadmium and zinc resistance from *E. coli*)
	Source organism	*E. coli*
	Reporter gene	*luxCDABE* (*V. f.*)
	Reporter design	Two plasmid system: pSLZntR (*zntR*) and pDNPzntAlux (*zntAp-lux*) or mini-Tn5(zntRPzntAlux)
	Host strain for construction	*E. coli* MC1061 or *P. fluorescens* OS8
	Assay	Fresh – expo phase, M9 medium with Glu, 30°C, 2 h assay time
	Measurement	Luminometry
	Detected compounds	$CdCl_2$, $HgCl_2$, $Pb(NO_3)_2$, $ZnSO_4$
	Concentration range	10 nM – 1 µM
	Application	Aqueous solutions
	Reference	(Ivask, Rolova et al. 2009)
Chromium	Switch	P_{chrB} (*chrB*)
	Source genes	*chrBA* (chromate resistance)
	Source organism	*C. metallidurens* plasmid pMOL28
	Reporter gene	*luxCDABE* (*V. f.*)
	Reporter design	Prom. fusion: plasmid: pEBZ141 (P_{chrB}-*chrBA'*::*luxCDABE*) Regulator: chromosome: host encoded
	Host strain for construction	*C. metallidurens* – Homologous - Resistance: +
	Assay	Fresh, MM glycerol, 28°C, 120 min assay time
	Measurement	Luminometry
	Detected compounds	CrO_4^{2-}, $Cr_2O_4^{2-}$, Cr^{3+}
	Concentration range	1 nM - 0.1 mM (50 mM)
	Application	Slurries or aqueous solutions
	Reference	(Peitzsch, Eberz et al. 1998)
Chromium	Switch	P_{chrB} (*chrB*)
	Source genes	*chrBA* (chromate resistance)
	Source organism	*C. metallidurens* plasmid pMOL28
	Reporter gene	*luxCDABE* (*V. f.*)
	Reporter design	Prom. fusion + regulator: plasmid: not further described plasmid (P_{chrB}-*chrBA'*::*luxCDABE*) Regulator: chromosome: host encoded

Compound	Nature	Specificities
	Host strain for construction	*C. metallidurens* – Homologous - Resistance: +
	Assay	Fresh, MM gluconate, 23°C, 300 min assay time
	Measurement	Luminometry
	Detected compounds	CrO_4^{2-}, Cr^{3+}
	Concentration range	2 µM - 80 µM (40 µM)
	Application	Slurries or aqueous solutions
	Reference	(Corbisier, van der Lelie et al. 1999)
Copper	Switch	ND - *cup* genes
	Source genes	*cupC* (copper resistance)
	Source organism	*C. metallidurens* plasmid pMOL28
	Reporter gene	*luxCDABE* (V. f.)
	Reporter design	Prom. fusion + regulator: plasmid: derivative of plasmid pMOL90, containing transposon insertion in *cupC* gene (*cupC'*-Tn[*luxCDABE*])
	Host strain for construction	*C. metallidurens* – Homologous - Resistance: +
	Assay	Lyophilized, MM acetate, 23°C, 300 min assay time
	Measurement	Luminometry
	Detected compounds	Cu^{2+}, Cd^{2+}, Zn^{2+}
	Concentration range	2 µM - 0.1 mM (40 µM)
	Application	Fly-ashes (Corbisier, Thiry et al. 1996), heavy metal-contaminated soil slurries (Corbisier, Thiry et al. 1996)
	Reference	(Corbisier, Thiry et al. 1994)[n]
Copper	Switch	P_{copA} - CueR
	Source genes	*copA*, *cueR* (copper tolerance)
	Source organism	*E. coli*
	Reporter gene	*luxCDABE (P. l.)* or *lucFF(P. p.)*
	Reporter design	Two plasmid system: pSLcueR (*zntR*) and pDNPcopAlux (*zntAp-lux*) or mini-Tn5(cueRPcopAlux)
	Host strain for construction	*E. coli* MC1061 or *P. fluorescens* OS8
	Assay	Fresh – expo phase, M9 medium with Glu, 30°C, 2 h assay time
	Measurement	Luminometry
	Detected compounds	$CuSO_4$, $AgNO_3$,
	Concentration range	100 nM – 1 mM
	Application	Aqueous solutions
	Reference	(Rensing, Fan et al. 2000; Hakkila, Green et al. 2004; Ivask, Rolova et al. 2009)
Lead	Switch	P_{pbrA} - PbrR
	Source genes	*pbrRA* (lead resistance)
	Source organism	*C. metallidurens* plasmid pMOL30
	Reporter gene	*luxCDABE (P. l.)*
	Reporter design	Prom fusion + regulator: pDNpbRPpbrAlux or mini-Tn5(pbrRPpbrAlux)
	Host strain for construction	*P. fluorescens* OS8 Heterologous Resistance:

Compound	Nature	Specificities
	Assay	Fresh – expo phase, M9 medium with Glu, 30°C, 2 h assay time
	Measurement	Luminometry
	Detected compounds	Pb^{2+}, Cd^{2+}, Hg^{2+}, Zn^{2+},
	Concentration range	0.2 – 1 mM
	Application	Aqueous solutions
	Reference	(Ivask, Rolova et al. 2009)
Lead	Switch	P_{pbrA} - PbrR
	Source genes	*pbrRA* (lead resistance)
	Source organism	*C. metallidurens* plasmid pMOL30
	Reporter gene	*luxCDABE* (V. f.)
	Reporter design	Prom. fusion + regulator: plasmid: not further described plasmid (P_{pbrR}-*pbrRA'::luxCDABE*)
	Host strain for construction	*C. metallidurens* – Homologous - Resistance: +
	Assay	Lyophilized, MM gluconate, 23°C, 300 min assay time
	Measurement	Luminometry
	Detected compounds	Pb^{2+}
	Concentration range	0.5 mM - 20 mM (5 mM)
	Application	Aqueous suspensions
	Reference	(Corbisier, van der Lelie et al. 1999)
Mercury	Switch	P_{TPCAD} (*merT*) – MerR (activator/repressor)
	Source genes	*merTPCAD* (mercury resistance)
	Source organism	*E. coli* transposon Tn*21*
	Reporter gene	*luxAB* (V. h.)
	Reporter design	Prom. fusion + regulator: plasmid: pCC306 (*merR*- P_{TPCAD} ::*luxAB*)
	Host strain for construction	*E. coli* - Homologous - Resistance: -
	Assay	Fresh, rich medium, 37°C, 15 min assay time
	Measurement	Luminometry
	Detected compounds	Hg^{2+}
	Concentration range	20 nM - 2 mM (0.2 mM)
	Application	Aqueous solutions
	Reference	(Condee and Summers 1992)
Mercury	Switch	P_{TPCAD} (*merT*) – MerR (activator/repressor)
	Source genes	*merTPCAD* (mercury resistance)
	Source organism	*E. coli* transposon Tn*21*
	Reporter gene	*luxCDABE* (V. f.)
	Reporter design	Prom. fusion + regulator: plasmid: pRB28 (*merR*-P_{TPCAD}::*luxCDABE*)
	Host strain for construction	*E. coli* - Homologous - Resistance: -
	Assay	Fresh, MM pyruvate, RT, 100 min assay time
	Measurement	
	Detected compounds	Hg^{2+}
	Concentration range	1 nM - 1 µM (0.5 mM)

Compound	Nature	Specificities
	Application	mercury-contaminated surface water, aqueous soil extract of mercury contaminated soil (Rasmussen, Sørensen et al. 2000)
	Reference	(Selifonova, Burlage et al. 1993)
Mercury	Switch	P_{TPCAD} (*merT*) – MerR (activator/repressor)
	Source genes	*merTPCAD* (mercury resistance)
	Source organism	*E. coli* transposon Tn*21*
	Reporter gene	*luxCDABE* (*V. f.*)
	Reporter design	<u>Prom. fusion + regulator</u>: plasmid: pOS14 (*merR*-P$_{TPCAD}$-*merTPC::luxCDABE*)
	Host strain for construction	*E. coli* - Homologous - Resistance: +
	Assay	Fresh, MM pyruvate, RT, 100 min assay time
	Measurement	Luminometry
	Detected compounds	Hg^{2+}
	Concentration range	0.5 nM - 0.5 µM (50 nM)
	Application	mercury-contaminated surface water
	Reference	(Selifonova, Burlage et al. 1993)
Mercury	Switch	P_{TPCAD} (*merT*) – MerR (activator/repressor)
	Source genes	*merTPCAD* (mercury resistance)
	Source organism	*E. coli* transposon Tn*21*
	Reporter gene	*lucFF* (*P. p.*)
	Reporter design	<u>Prom. fusion + regulator</u>: plasmid: pTOO11 (*merR*-P$_{TPCAD}$::*lucFF*)
	Host strain for construction	*E. coli* - Homologous - Resistance: -
	Assay	Fresh, MM casein, 30°C, 60 min assay time
	Measurement	Luminometry
	Detected compounds	Hg^{2+}, Cd^{2+}
	Concentration range	0.1 fM - 1 mM (0.1 mM)
	Application	Aqueous solutions
	Reference	(Virta, Lampinen et al. 1995)
Mercury	Switch	P_{TPCAD} (*merT*) – MerR (activator/repressor)
	Source genes	*merTPCAD* (mercury resistance)
	Source organism	*E. coli* transposon Tn*21*
	Reporter gene	*luxCDABE* (V.f.) or Wt *gfp* (*A. v.*)
	Reporter design	<u>Prom. fusion + regulator</u>: plasmid: pUT-*mergfp* (Tn[*merR*-P$_{mer}$::*gfp*])
	Host strain for construction	*E. coli* - Homologous - Resistance: -
	Assay	Fresh, MM leucine + proline, RT, 80 min assay time or fresh, rich medium, 30°C, 16 h assay time
	Measurement	Luminometry or fluorometry
	Detected compounds	Hg^{2+}
	Concentration range	0.25 - 2.5 ng/ml (as $HgCl_2$) or 50 - 250 ng/ml for gfp
	Application	Aqueous soil extracts[v]
	Reference	(Hansen and Sørensen 2000)

Compound	Nature	Specificities
Mercury	Switch	P*TPCAD* (*merT*) – MerR (activator/repressor)
	Source genes	*merTPCAD* (mercury resistance)
	Source organism	*E. coli* transposon Tn*21*
	Reporter gene	*lucFF* or *gfp* or *dsred* or *luxCDABE*
	Reporter design	Prom. fusion + regulator: plasmid: pmerRlucFFGFP, pmerRGFP, pmerRDsred, pmerRluxCDABE
	Host strain for construction	*E. coli* - Homologous - Resistance: +
	Assay	Fresh, M9 plus casein, 37°C, 0.5- 4 h assay time
	Measurement	Luminometry or Fluorimetry
	Detected compounds	Hg
	Concentration range	100 pM – 2 nM (*luc or lux*)
	Application	Aqueous solutions
	Reference	(Hakkila, Maksimow et al. 2002)
Mercury	Switch	P*merT* (*merT*) – MerR (activator/repressor)
	Source genes	*merTA* (mercury resistance)
	Source organism	*Serratia marcescens*
	Reporter gene	*luxCDABE* (*V. f.*)
	Reporter design	Prom. fusion + regulator: plasmid: pMerlux (not further described)
	Host strain for construction	*E. coli* - Heterologous - Resistance: -
	Assay	Fresh, rich medium, 28°C, 30 min assay time
	Measurement	Luminometry
	Detected compounds	Hg^{2+}, mercuric acetate, Hg^+
	Concentration range	20 nM - 4 mM (0.4 mM)
	Application	Aqueous solutions
	Reference	(Tescione and Belfort 1993)
Mercury	Switch	P*merT* (*merT*) – MerR (activator/repressor)
	Source genes	*merTA* (mercury resistance)
	Source organism	*Serratia marcescens*
	Reporter gene	*luxCDABE* (*V. f.*)
	Reporter design	Prom. fusion + regulator: plasmid pmerR*BS*Bpmerlux or pDNmerR*BS*Bpmerlux or mini-Tn5(merR*BS*Bpmerlux)
	Host strain for construction	*E. coli* MC1061 or *P. fluorescens* OS8
	Assay	Fresh – expo phase, M9 medium with Glu, 30°C, 2 h assay time
	Measurement	Luminometry
	Detected compounds	CH_3HgCl, $HgCl_2$, $CdCl_2$
	Concentration range	8·pM – 1 nM
	Application	mercury-contaminated surface water
	Reference	(Ivask, Hakkila et al. 2001; Ivask, Rolova et al. 2009)
Nickel	Switch	*cnrHp* (*cnrH*) CnrY, CnrX (anti ECF sigma-factor complex)
	Source genes	*cnrHCBA* (cobalt/nickel resistance)
	Source organism	*C. metallidurens* plasmid pMOL28

Compound	Nature	Specificities
	Reporter gene	*luxCDABE* (*V. f.*)
	Reporter design	Prom. fusion + regulator: plasmid: pMOL1550 (*cnrYXH::luxCDABE*)
	Host strain for construction	*C. metallidurens* (pMOL28, pMOL30) - Homologous - Resistance: +
	Assay	Fresh, MM gluconate, 23°C, 16 h assay time
	Measurement	Luminometry
	Detected compounds	Ni^{2+}, Co^{2+}
	Concentration range	20 µM - 1 mM (0.3 mM)
	Application	Contaminated groundwaters or soil-water extracts
	Reference	(Tibazarwa, Wuertz et al. 2000)
Nickel	Switch	P_{cel} (*celA*)
	Source genes	*celABCDF* (cellobiose degradation)*i*
	Source organism	*E. coli*
	Reporter gene	*luxAB* (*V. h.*)
	Reporter design	Prom. fusion: chromosome: transposon insertion in *celF* gene (P_{cel}-*celABCDF'*-Tn[*luxAB*]) Regulator: chromosome: host encoded°
	Host strain for construction	*E. coli* – Homologous - Resistance: -
	Assay	Fresh, rich medium, 37°C, 18-24 h assay time
	Measurement	Luminometry
	Detected compounds	Ni^{2+}, Co^{2+}
	Concentration range	4 µM – 0.2 mM
	Application	Aqueous solution
	Reference	(Guzzo and DuBow 1994)
Thallium	Switch	ND
	Source genes	*tll* genes (thallium resistance)
	Source organism	*C. metallidurens* plasmid pMOL30
	Reporter gene	*luxCDABE* (*V. f.*)
	Reporter design	Prom. fusion + regulator: plasmid: derivative of plasmid pMOL30, containing transposon insertion in *tllB* gene (*tllB'*-Tn[*luxCDABE*])
	Host strain for construction	*C. metallidurens* – Homologous – Resistance: +
	Assay	No details
	Measurement	Luminometry
	Detected compounds	Tl^+
	Concentration range	4 µM - 0.1 mM (40 µM)
	Application	Aqueous solution
	Reference	(Collard, Corbisier et al. 1994)
Zinc	Switch	P_{czcD} (*czcD*) CzcI/CzcD (putative response regulator/periplasmic sensor protein)
	Source genes	*czcDRS* (regulation of the *czcCBA* genes encoding cobalt, zinc and cadmium resistance)
	Source organism	*C. metallidurens* plasmid pMOl30
	Reporter gene	*luxCDABE* (*V. f.*)

Compound	Nature	Specificities
	Reporter design	Prom. fusion + regulator: plasmid: derivative of plasmid pMOL30 containing transposon insertion in *czcS* gene (*czcICBADRS'*-Tn[*luxCDABE*])
	Host strain for construction	*C. metallidurens* – Homologous – Resistance: +
	Assay	Fresh, MM gluconate, 25°C, 48 h assay time
	Measurement	Luminometry
	Detected compounds	Zn^{2+}, Pb^{2+}, Cd^{2+}, Co^{2+}
	Concentration range	0.1 mM - 0.4 mM (0.2 mM)
	Application	Aqueous solution
	Reference	(van der Lelie, Schwuchow et al. 1997)
Zinc	Switch	P$_{zntA}$ - ZntR
	Source genes	*zntRA* (cadmium and zinc resistance)
	Source organism	*E. coli*
	Reporter gene	*luxCDABE (P. l.)*
	Reporter design	Two plasmid system: pSLZntR (*zntR*) and pDNPzntAlux (*zntAp-lux*) or mini-Tn5(zntRPzntAlux)
	Host strain for construction	*E. coli* MC1061 or *P. fluorescens* OS8
	Assay	Fresh – expo phase, M9 medium with Glu, 30°C, 2 h assay time
	Measurement	Luminometry
	Detected compounds	CdCl$_2$, HgCl$_2$, Pb(NO$_3$)$_2$, ZnSO$_4$
	Concentration range	1 µM – 1 mM
	Application	Aqueous solutions
	Reference	(Ivask, Rolova et al. 2009)
Zinc	Switch	P$_{smtA}$ (*smtA*) SmtB (repressor)
	Source genes	*smtA* (metallothionein conferring resistance against zinc)
	Source organism	*Synechococcus* sp.
	Reporter gene	*luxCDABE (V. f.)*
	Reporter design	Prom. fusion: plasmid: pJLE23 (P$_{smtA}$::*luxCDABE*) Regulator: chromosome: host encoded
	Host strain for construction	*Synechococcus* sp. - Homologous- Resistance: +
	Assay	Fresh, MM, 30°C, 60 min assay time
	Measurement	Luminometry
	Detected compounds	Zn^{2+}, Cd^{2+}, Cu^{2+}
	Concentration range	0.5 mM - 16 mM (4 mM)
	Application	Aqueous solution
	Reference	(Erbe, Adams et al. 1996)

REFERENCES

Baumann, B. and J. R. van der Meer (2007). "Analysis of bioavailable arsenic in rice with whole cell living bioreporter bacteria." *J. Agric. Food Chem.* **55**(6): 2115–2120. DOI: 10.1021/jf0631676

Cai, J. and M. S. DuBow (1996). "Expression of the *Escherichia coli* chromosomal *ars* operon." *Can. J. Microbiol.* **42**: 662–671.

Cai, J. and M. S. DuBow (1997). "Use of a luminescent bacterial biosensor for biomonitoring and characterization of arsenic toxicity of chromated copper arsenate (CCA)." *Biodegradation* **8**: 105–111. DOI: 10.1023/A:1008281028594

Collard, J. M., P. Corbisier, et al. (1994). "Plasmids for heavy metal resistance in *Alcaligenes eutrophus* CH34: mechanisms and applications." *FEMS Microbiol. Rev.* **14**(4): 405–414. DOI: 10.1111/j.1574-6976.1994.tb00115.x

Condee, C. W. and A. O. Summers (1992). "A *mer-lux* transcriptional fusion for real-time examination of *in vivo* gene expression kinetics and promoter response to altered superhelicity." *J. Bacteriol.* **174**(24): 8094–8101.

Corbisier, P., G. Ji, et al. (1993). "*luxAB* gene fusions with the arsenic and cadmium resistance operons of *Staphylococcus aureus* plasmid pI258." *FEMS Microbiol. Lett.* **110**: 231–238. DOI: 10.1111/j.1574-6968.1993.tb06325.x

Corbisier, P., E. Thiry, et al. (1996). "Bacterial biosensors for the toxicity assessment of solid wastes." *Environ. Tox. Wat. Qual.* **11**(3): 171–177. DOI: 10.1002/(SICI)1098-2256(1996)11:3%3C171::AID-TOX1%3E3.0.CO;2-6

Corbisier, P., E. Thiry, et al. (1994). Construction and development of metal ion biosensors. *Bioluminescence and Chemiluminescence: Fundamentals and Applied Aspects*. A. K. Campbell, L. J. Kricka and P. E. Stanley. Chichester, New York, Brisbane, Toronto, Singapore, John Wiley & Sons: 151–155.

Corbisier, P., D. van der Lelie, et al. (1999). "Whole cell- and protein-based biosensors for the detection of bioavailable heavy metals in environmental samples." *Anal. Chim. Acta* **387**: 235–244. DOI: 10.1016/S0003-2670(98)00725-9

Erbe, J. L., A. C. Adams, et al. (1996). "Cyanobacteria carrying an *smt-lux* transcriptional fusion as biosensors for the detection of heavy metal cations." *J. Ind. Microbiol.* **17**(2): 80–83. DOI: 10.1007/BF01570047

Guzzo, A. and M. S. DuBow (1994). "A *luxAB* transcriptional fusion to the cryptic *celF* gene of *Escherichia coli* displays increased luminescence in the presence of nickel." *Mol. Gen. Genet.* **242**: 455–460. DOI: 10.1007/BF00281796

Hakkila, K., T. Green, et al. (2004). "Detection of bioavailable heavy metals in EILATox-Oregon samples using whole-cell luminescent bacterial sensors in suspension or immobilized onto fibre-optic tips." *J. Appl. Toxicol.* **24**(5): 333–342. DOI: 10.1002/jat.1020

Hakkila, K., M. Maksimow, et al. (2002). "Reporter genes *lucFF*, *luxCDABE*, *gfp*, and *dsred* have different characteristics in whole-cell bacterial sensors." *Anal. Biochem.* **301**: 235–242. DOI: 10.1006/abio.2001.5517

Hansen, L. H. and S. J. Sørensen (2000). "Versatile biosensor vectors for detection and quantification of mercury." *FEMS Microbiol. Lett.* **193**: 123–127. DOI: 10.1111/j.1574-6968.2000.tb09413.x

Ivask, A., K. Hakkila, et al. (2001). "Detection of organomercurials with sensor bacteria." *Anal. Chem.* **73**(21): 5168–5171. DOI: 10.1021/ac010550v

Ivask, A., T. Rolova, et al. (2009). "A suite of recombinant luminescent bacterial strains for the quantification of bioavailable heavy metals and toxicity testing." *BMC Biotechnol.* **9**: 41. DOI: 10.1186/1472-6750-9-41

Liao, V. H., M. T. Chien, et al. (2006). "Assessment of heavy metal bioavailability in contaminated sediments and soils using green fluorescent protein-based bacterial biosensors." *Environ. Pollut.* **142**(1): 17–23. DOI: 10.1016/j.envpol.2005.09.021

Peitzsch, N., G. Eberz, et al. (1998). "*Alcaligenes eutrophus* as a bacterial chromate sensor." *Appl. Environ. Microbiol.* **64**(2): 453–458.

Ramanathan, S., W. Shi, et al. (1997). "Sensing antimonite and arsenite at the subattomole level with genetically engineered bioluminescent bacteria." *Anal. Chem.* **69**: 3380–3384. DOI: 10.1021/ac970111p

Rasmussen, L. D., S. J. Sørensen, et al. (2000). "Application of a *mer-lux* biosensor for estimating bioavailable mercury in soil." *Soil Biol. Biochem.* **32**: 639–646. DOI: 10.1016/S0038-0717(99)00190-X

Rensing, C., B. Fan, et al. (2000). "CopA: An *Escherichia coli* Cu(I)-translocating P-type ATPase." *Proc. Natl. Acad. Sci. USA* **97**(2): 652–656.

Selifonova, O., R. Burlage, et al. (1993). "Bioluminescent sensors for detection of bioavailable Hg(II) in the environment." *Appl. Environ. Microbiol.* **59**(9): 3083–3090.

Stocker, J., D. Balluch, et al. (2003). "Development of a set of simple bacterial biosensors for quantitative and rapid field measurements of arsenite and arsenate in potable water." *Environ. Sci. Technol.* **37**: 4743–4750. DOI: 10.1021/es034258b

Tauriainen, S., M. Karp, et al. (1997). "Recombinant luminescent bacteria for measuring bioavailable arsenite and antimonite." *Appl. Environ. Microbiol.* **63**(11): 4456–4461.

Tauriainen, S., M. Karp, et al. (1998). "Luminescent bacterial sensor for cadmium and lead." *Biosens. Bioelectron.* **13**: 931–938. DOI: 10.1016/S0956-5663(98)00027-X

Tauriainen, S., M. Virta, et al. (1999). "Measurement of firefly luciferase reporter gene activity from cells and lysates using *Escherichia coli* arsenite and mercury sensors." *Anal. Biochem.* **272**: 191–198. DOI: 10.1006/abio.1999.4193

Tescione, L. and G. Belfort (1993). "Construction and evaluation of a metal ion biosensor." *Biotechnol. Bioeng.* **42**(8): 945–952. DOI: 10.1002/bit.260420805

Tibazarwa, C., S. Wuertz, et al. (2000). "Regulation of the *cnr* cobalt and nickel resistance determinant of *Ralstonia eutropha* (*Alcaligenes eutrophus*) CH34." *J. Bacteriol.* **182**(5): 1399–1409. DOI: 10.1128/JB.182.5.1399-1409.2000

van der Lelie, D., T. Schwuchow, et al. (1997). "Two-component regulatory system involved in transcriptional control of heavy-metal homeostasis in *Alcaligenes eutrophus*." *Mol. Microbiol.* **23**(3): 493–503. DOI: 10.1046/j.1365-2958.1997.d01-1866.x

Virta, M., J. Lampinen, et al. (1995). "A luminescence-based mercury biosensor." *Anal. Chem.* **67**(3): 667–669. DOI: 10.1021/ac00099a027

Wackwitz, A., H. Harms, et al. (2008). "Internal arsenite bioassay calibration using multiple reporter cell lines." *Microb. Biotechnol.* **1**(2): 149–157. DOI: 10.1111/j.1751-7915.2007.00011.x

APPENDIX C

Bacterial Bioreporter Designs Responsive to Toxicity or Stress Conditions

Mechanism	Characteristic	Specificities
alkA DNA alkylation	Switch	P*alkA* (*alkA*) Ada (activator)
	Source genes	*alkA* (DNA repair enzyme able to excise methylated bases from DNA)
	Source organism	*E. coli*
	Reporter gene	*luxCDABE* (*V. f.*)
	Reporter design	Prom. fus.: plasmid: pAlkALux1 (P*alkA*::*luxCDABE*) Regulator: chromosome: host encoded
	Host strain for construction	*E. coli* – Homologous
	Assay	Lyophilized, MM + casein, 30°C, 90 min incubation time
	Measurement	Luminometry
	Detected compounds	MNNG[i]
	Concentration range	
	Stress	Methylated phosphotriesters and bases generated by DNA alkylation
	Reference	(Vollmer, Belkin et al. 1997)
fabA Fatty acid biosynthesis	Switch	P*fabA* (*fabA*) FadR (activator)
	Source genes	Beta-hydroxy-decanoyl-ACP-dehydrase
	Source organism	*E. coli* chromosome
	Reporter gene	*luxCDABE* (*V. f.*)
	Reporter design	Prom. fusion: plasmid pFabA'-Lux Regulator host chromosome encoded
	Host strain for construction	*E. coli rpsL200, tolC*
	Assay	Fresh, rich medium, 37°C, 60 min induction time
	Measurement	Luminometry
	Detected compounds	Phenol
	Concentration range	
	Stress	Balance between fatty acid synthesis and degradation
	Reference	(Ben-Israel, Ben-Israel et al. 1998; Bechor, Smulski et al. 2002; Pedahzur, Polyak et al. 2004)
clpB Heat shock	Switch	P*clpB* (*clpB*) DnaK (anti-s^H factor)
	Source genes	*dnaK* (chaperone; anti-s^H factor)
	Source organism	*E. coli*
	Reporter gene	*gfpuv* (*A. v.*)
	Reporter design	Prom. fus.: plasmid: pGFPuv-ClpB (P*clpB*::*gfp*uv) Regulator: chromosome; host encoded
	Host strain for construction	*E. coli* - Homologous
	Assay	
	Measurement	Fluorometry

Mechanism	Characteristic	Specificities
	Detected compounds	Phenol
	Concentration range	
	Stress	Increase in the level of misfolded proteins due to the presence of certain compounds or a raise in temperature
	Reference	(Cha, Srivastava et al. 1999)
dnaK Heat shock	Switch	P_{dnaK} (*dnaK*) DnaK (anti-s^H factor)
	Source genes	*dnaK* (chaperone; anti-s^H factor)
	Source organism	*E. coli*
	Reporter gene	*luxCDABE* (*V. f.*)
	Reporter design	Prom. fus.: plasmid: pRY002 (P_{dnaK}::*luxCDABE*) Regulator: chromosome: host encoded
	Host strain for construction	*E. coli* - homologous
	Assay	
	Measurement	Luminometry
	Detected compounds	pentachlorophenol, 2,4-dichlorophenoxyacetic acid, 2-/4-nitrophenol, phenol, Cu^{2+}
	Concentration range	
	Application	
	Reference	(Van Dyk, Majarian et al. 1994)
dnaK Heat shock	Switch	P_{dnaK} (*dnaK*) DnaK (anti-s^H factor)
	Source genes	*dnaK* (chaperone; anti-s^H factor)
	Source organism	*E. coli*
	Reporter gene	*Gfp-uv*
	Reporter design	Prom. fus.: plasmid: pGFPuv-DnaK (P_{dnaK}::*gfp_{uv}*) Regulator: chromosome: host encoded
	Host strain for construction	*E. coli* – Homologous
	Assay	
	Measurement	Fluorometry
	Detected compounds	Phenol
	Concentration range	
	Application	
	Reference	(Cha, Srivastava et al. 1999)
grpE heat shock	Switch	P_{grpE} (*grpE*) DnaK (anti-s^H factor)
	Source genes	*grpE* (chaperone) (Yura and Nakahigashi 1999)
	Source organism	*E. coli*
	Reporter gene	*luxCDABE* (*V. f.*)
	Reporter design	Prom. fus.: plasmid: pGrpELux5 (P_{grpE}::*luxCDABE*) Regulator: chromosome: host encoded
	Host strain for construction	*E. coli* – Homologous
	Assay	
	Measurement	Luminometry

Mechanism	Characteristic	Specificities
	Detected compounds	pentachlorophenol, 2,4-dichlorophenoxyacetic acid, 2-/4-nitrophenol, phenol, Cu^{2+}, 2,4-dinitrophenol, Cd^{2+} (Van Dyk, Smulski et al. 1995), 2-chlorophenol, 4-bromophenol, methylene chloride, methylene bromid, chloroform, bromoform, bromodichloromethane (Belkin, Smulski et al. 1997)
	Concentration range	
	Application	
	Reference	(Van Dyk, Majarian et al. 1994)
lon Heat shock	Switch	P_{lon} (*lon*) DnaK (anti-s^H factor)
	Source genes	*lon* (protease)
	Source organism	*E. coli*
	Reporter gene	*luxCDABE* (*V. f.*)
	Reporter design	Prom. fus.: plasmid: pLonLux2 (P_{lon}::*luxCDABE*) Regulator: chromosome: host encoded
	Host strain for construction	*E. coli*
	Assay	
	Measurement	Luminometry
	Detected compounds	Pentachlorophenol
	Concentration range	
	Application	
	Reference	(van Dyk, Reed et al. 1995)
rpoH Heat shock	Switch	P_{rpoH} (*rpoH*) DnaK (anti-s^H factor)
	Source genes	*rpoH* sigma factor
	Source organism	*E. coli*
	Reporter gene	*Gfp-uv*
	Reporter design	Prom. fus.: plasmid: pGFPuv-Sigma (P_{rpoH}::*gfp_{uv}*) Regulator: chromosome: host encoded
	Host strain for construction	*E. coli* - homologous
	Assay	
	Measurement	Fluorometry
	Detected compounds	phenol
	Concentration range	
	Application	
	Reference	(Cha, Srivastava et al. 1999)
katG oxidative stress	Switch	P_{katG} (*katG*) OxyR (activator)
	Source genes	*katG* (catalase) (Storz and Imlay 1999)
	Source organism	*E. coli*
	Reporter gene	*luxCDABE* (*V. f.*)
	Reporter design	Prom. fus.:plasmid: pKatGLux2 (P_{katG}::*luxCDABE*) Regulator: chromosome: host encoded
	Host strain for construction	*E. coli* - Homologous
	Assay	

Mechanism	Characteristic	Specificities
	Measurement	Luminometry
	Detected compounds	catechol (Schweigert, Belkin et al. 1999), phenol, 2-/4-nitrophenol, 2-chlorophenol, 4-bromophenol (Belkin, Smulski et al. 1997), Hg^{2+}, Pb^{2+} ,Zn^{2+} (Ben-Israel, Ben-Israel et al. 1998)
	Concentration range	
	Stress	H$_2$O$_2$, nitrosothiols
	Reference	(Belkin, Smulski et al. 1996)
micF odixative stress	Switch	P$_{micF}$ (*micF*) SoxR/SoxS (activators)(Storz and Imlay 1999); MarA (Oh, Cajal et al. 2000)
	Source genes	*micF* (antisense RNA against *ompF*)
	Source organism	*E. coli*
	Reporter gene	*luxCDABE* (*V.f.*)
	Reporter design	Prom. fus.: plasmid: pMicFLux1 (P$_{micF}$::*luxCDABE*) Regulator: chromosome: host encoded
	Host strain for construction	*E. coli* - Homologous
	Assay	
	Measurement	Luminometry
	Detected compounds	paraquat, 4-nitrophenol, 2-chlorophenol, 4-bromophenol, bromoform, bromodichloromethane (Belkin, Smulski et al. 1997)
	Concentration range	
	Stress	Superoxide anion (O$_2^{-}$), nitric oxide (NO) (Storz and Imlay 1999)
	Reference	(Dukan, Dadon et al. 1996; Oh, Cajal et al. 2000)
fumC oxidative stress	Switch	P$_{fumC}$ (*fumC*) - SoxR/SoxS
	Source genes	*fumC* fumarase C
	Source organism	*E. coli* RFM443 chromosome
	Reporter gene	*luxCDABE* (*V. f.*)
	Reporter design	Prom fusion: plasmid pBC-Fum-Lux
	Host strain for construction	E. coli *RFM443*
	Assay	
	Measurement	Luminometry
	Detected compounds	paraquat
	Concentration range	
	Application	
	Reference	(Kim, Park et al. 2005)
soxS oxidative stress	Switch	P$_{soxS}$ (*soxS*)- SoxR
	Source genes	*soxS* dual transcription activator in superoxide response regulon
	Source organism	*E. coli* RFM443 chromosome
	Reporter gene	*luxCDABE* (*V. f.*)
	Reporter design	Prom fusion: plasmid pBC-Sox-Lux
	Host strain for construction	*E. coli* RFM443
	Assay	

Mechanism	Characteristic	Specificities
	Measurement	Luminometry
	Detected compounds	<u>paraquat</u>
	Concentration range	
	Application	
	Reference	(Kim, Youn et al. 2005)
inaA oxidative stress	Switch	P*inaA*(*inaA*) - SoxR/SoxS
	Source genes	*inaA* response to acid conditions
	Source organism	*E. coli* RFM443 chromosome
	Reporter gene	*luxCDABE* (*V. f.*)
	Reporter design	<u>Prom fusion</u>: plasmid pBC-Ina-Lux
	Host strain for construction	*E. coli* RFM443
	Assay	
	Measurement	Luminometry
	Detected compounds	<u>paraquat</u>
	Concentration range	
	Application	
	Reference	(Kim, Youn et al. 2005)
sodA oxidative stress	Switch	P*sodA* (*sodA*) - SoxR/SoxS
	Source genes	*sodA* superoxide dismutase
	Source organism	*E. coli* RFM443 chromosome
	Reporter gene	*luxCDABE* (*V. f.*)
	Reporter design	<u>Prom fusion</u>: plasmid pBC-SodA-Lux
	Host strain for construction	*E. coli* RFM443
	Assay	
	Measurement	Luminometry
	Detected compounds	<u>paraquat</u>
	Concentration range	
	Application	
	Reference	(Kim, Youn et al. 2005; Lee, Youn et al. 2007)
cda SOS response	Switch	P*cda* (*cda*) LexA/RecA (repressor/anti-repressor)
	Source genes	*cda* (protein synthesis inhibitor)
	Source organism	*E. coli* plasmid pColD
	Reporter gene	*luxCDABE* (*P. l.*)
	Reporter design	<u>Prom. fus.</u>: plasmid: pPLS-1 (P*cda*::*luxCDABFE*) <u>Regulator</u>: chromosome: host encoded
	Host strain for construction	*E. coli*
	Assay	
	Measurement	Luminometry
	Detected compounds	dimethylsulfate, $Cr_2O_7^{2-}$ (Rettberg, Baumstark-Khan et al. 1999)
	Concentration range	

Mechanism	Characteristic	Specificities
	Stress	Single stranded DNA strands that appear as a consequence of DNA replication inhibition
	Reference	(Ptitsyn, Horneck et al. 1997)
recA SOS response	Switch	P_{recA} - LexA/RecA (repressor/anti-repressor)
	Source genes	*recA* recombinase involved in post-replication DNA repair
	Source organism	*E. coli*
	Reporter gene	*gfp-mut3 (A. v.)*
	Reporter design	Prom. fus.: plasmid: pRGM5 (P_{recA}::*gfp*-mut3) Regulator: chromosome: host encoded
	Host strain for construction	*E. coli*
	Assay	
	Measurement	Fluorometry
	Detected compounds	MNNG, nalidixic acid, mitomycin C
	Concentration range	
	Application	
	Reference	(Kostrzynska, Leung et al. 2002)
recA SOS response	Switch	P_{recA} - LexA/RecA (repressor/anti-repressor)
	Source genes	*recA* recombinase involved in post-replication DNA repair
	Source organism	*E. coli*
	Reporter gene	*luxCDABE (V. f.)*
	Reporter design	Prom. fus.: plasmid: pRecALux3 (P_{recA}::*luxCDABE*) Regulator: chromosome: host encoded
	Host strain for construction	*E. coli* - Homologous
	Assay	
	Measurement	Luminometry
	Detected compounds	Phenol, 2-chloropheno, methylene chloride, methylene bromide, chloroform (Belkin, Smulski et al. 1997), benzo[a]pyrene, naphthalene, Cd^{2+} (Min, Kim et al. 1999)
	Concentration range	
	Application	
	Reference	(Vollmer, Belkin et al. 1997)
recA SOS response	Switch	P_{recA} - LexA/RecA (repressor/anti-repressor)
	Source genes	*recA* recombinase involved in post-replication DNA repair
	Source organism	*Acinetobacter baylyi* ADP1
	Reporter gene	*luxCDABE (P. l.)*
	Reporter design	Chromosomal integration: recA'-luxCDABE
	Host strain for construction	*Acinetobacter baylyi*
	Assay	3 h
	Measurement	Luminometry
	Detected compounds	MMC, Pyrene, Benz[a]-pyrene
	Concentration range	20 nM - 2 mM (0.2 mM)

Mechanism	Characteristic	Specificities
	Application	MDL for MMC and benzo[a]pyrene 1.5 nM and 0.4 nM, respectively
	Reference	(Song, Li et al. 2009)
recN SOS response	Switch	P_{recN} (*recN*)
	Source genes	*recN* (protein involved in post-replication DNA repair)
	Source organism	*E. coli*
	Reporter gene	*luxCDABE* (*V. f.*)
	Reporter design	Prom. fus.: plasmid: pMOL1067 (P_{recN}::*luxCDABE*) Regulator: chromosome: host encoded
	Host strain for construction	*Salmonella typhimurium* -Heterologous
	Assay	
	Measurement	
	Detected compounds	Mitomycin C, nalidixic acid, carbaryl, pentachlorophenol, SeO$_2$, K$_2$Cr$_2$O$_7$, benzo[a]pyrene, chrysene, pyrene, 2,4,5,7-tetranitro-9-fluorenone, 4-nitroquinoline-*N*-oxide, fluoranthene, phenanthrene, *N*-nitrosodiethylamine, 2-aminofluorene, MMS, hydrazine
	Concentration range	
	Application	
	Reference	(van der Lelie, Regniers et al. 1997)
Umu SOS response	Switch	P_{umuD} (*umuD*) LexA/RecA
	Source genes	*umuDC* (enzymes involved in mutagenic DNA repair)
	Source organism	*E. coli*
	Reporter gene	*luxAB* (*V. h.*) or *gfp*
	Reporter design	Prom. fusion: plasmid: pTJ10 (*umuDC'*::*luxAB*) or plasmid: pTJ*gfp* (*umuDC'*::*gfp*) Regulator: chromosome: host encoded
	Host strain for construction	*E. coli* - Homologous
	Assay	
	Measurement	Luminometry or fluorometry
	Detected compounds	MNNG, MMS
	Concentration range	
	Application	
	Reference	(Justus and Thomas 1998)
uvrA SOS response	Switch	P_{uvrA} (*uvrA*) LexA/RecA
	Source genes	*uvrA* (DNA repair enzyme involved in the excision of pyrimidine dimers and bulky lesions)
	Source organism	*E. coli*
	Reporter gene	*luxCDABE* (V.f.)
	Reporter design	Prom. fus.: plasmid: pUvrALux1 (P_{uvrA}::*luxCDABE*) Regulator: chromosome: host encoded
	Host strain for construction	*E. coli* - Homologous
	Assay	

Mechanism	Characteristic	Specificities
	Measurement	Luminometry
	Detected compounds	Mitomycin C
	Concentration range	
	Application	
	Reference	(Vollmer, Belkin et al. 1997)
Usp	Switch	P_{uspA} (*uspA*)
	Source genes	*uspA* universal stress protein
	Source organism	*E. coli*
	Reporter gene	*luxCDABE* (V.f.)
	Reporter design	Prom. fus.: plasmid: pUspALux2 (P_{uspA}::*luxCDABE*) Regulator: chromosome: host encoded
	Host strain for construction	*E. coli* - Homologous - Resistance: -
	Assay	
	Measurement	Luminometry
	Detected compounds	4-nitrophenol, 2,4-dinitrophenol, phenol, pentachlorophenol, Cu^{2+}
	Concentration range	
	Application	
	Reference	(Van Dyk, Smulski et al. 1995)
sulA	Switch	P_{sulA} (*sulA*) – LexA/RecA
SOS response	Source genes	*sulA* SOS cell division inhibitor
	Source organism	*E. coli*
	Reporter gene	*luxCDABE* (P.l.)
	Reporter design	Prom. fusion:plasmid: psulA::luxCDABE Regulator provided on chromosome of host
	Host strain for construction	*E. coli* - Homologous
	Assay	Fresh, LB broth, shaking at 37°C to early exponential growth phase (OD600 a 0.12). 2 h assay time.
	Measurement	Luminometry
	Detected compounds	Nalidixic acid
	Concentration range	1 – 10 mg/L
	Application	Aqueous solutions
	Reference	(Yagur-Kroll, Bilic et al. 2009) (Norman, Hestbjerg Hansen et al. 2005)

REFERENCES

Bechor, O., D. R. Smulski, et al. (2002). "Recombinant microorganisms as environmental biosensors: pollutants detection by *Escherichia coli* bearing *fabA'::lux* fusions." *J. Biotechnol.* **94**(1): 125–132. DOI: 10.1016/S0168-1656(01)00423-0

Belkin, S., D. R. Smulski, et al. (1997). "A panel of stress-responsive luminous bacteria for the detection of selected classes of toxicants." *Wat. Res.* **31**(12): 3009–3016. DOI: 10.1016/S0043-1354(97)00169-3

Belkin, S., D. R. Smulski, et al. (1996). "Oxidative stress detection with *Escherichia coli* harboring a *katG'::lux* fusion." *Appl. Environ. Microbiol.* **62**: 2252–2256.

Ben-Israel, O., H. Ben-Israel, et al. (1998). "Identification and quantification of toxic chemicals by use of *Escherichia coli* carrying *lux* genes fused to stress promoters." *Appl. Environ. Microbiol.* **64**(11): 4346–4352.

Cha, H. J., R. Srivastava, et al. (1999). "Green fluorescent protein as a noninvasive stress probe in resting *Escherichia coli* cells." *Appl. Environ. Microbiol.* **65**(2): 409–414.

Dukan, S., S. Dadon, et al. (1996). "Hypochlorous acid activates the heat shock and *soxRS* systems of *Escherichia coli*." *Appl. Environ. Microbiol.* **62**(11): 4003–4008.

Justus, T. and S. M. Thomas (1998). "Construction of a *umuC'-luxAB* plasmid for the detection of mutagenic DNA repair via luminescence." *Mutation Res.* **398**: 131–141. DOI: 10.1016/S0027-5107(97)00215-7

Kim, B. C., C. H. Youn, et al. (2005). "Screening of target-specific stress-responsive genes for the development of cell-based biosensors using a DNA microarray." *Anal. Chem.* **77**: 8020–8026. DOI: 10.1021/ac0514218

Kim, M. N., H. H. Park, et al. (2005). "Construction and comparison of *Escherichia coli* whole-cell biosensors capable of detecting aromatic compounds." *J. Microbiol. Meth.* **60**: 235–245. DOI: 10.1016/j.mimet.2004.09.018

Kostrzynska, M., K. T. Leung, et al. (2002). "Green fluorescent protein-based biosensor for detecting SOS-inducing activity of genotoxic compounds." *J. Microbiol. Methods* **48**: 43–51. DOI: 10.1016/S0167-7012(01)00335-9

Lee, J. H., C. H. Youn, et al. (2007). "An oxidative stress-specific bacterial cell array chip for toxicity analysis." *Biosens. Bioelectron.* **22**(9–10): 2223–2229. DOI: 10.1016/j.bios.2006.10.038

Min, J., E. J. Kim, et al. (1999). "Distinct responses of a *recA::luxCDABE Escherichia coli* strain to direct and indirect DNA damaging agents." *Mutation Res.* **442**: 61–68. DOI: 10.1016/S1383-5718(99)00059-5

Norman, A., L. Hestbjerg Hansen, et al. (2005). "Construction of a ColD *cda* promoter-based SOS-green fluorescent protein whole-cell biosensor with higher sensitivity toward genotoxic compounds than constructs based on *recA*, *umuDC*, or *sulA* promoters." *Appl. Environ. Microbiol.* **71**(5): 2338–2346. DOI: 10.1128/AEM.71.5.2338-2346.2005

Oh, J. T., Y. Cajal, et al. (2000). "Cationic peptide antimicrobials induce selective transcription of *micF* and *osmY* in *Escherichia coli*." *Biochim. Biophys. Acta* **1463**(1): 43–54. DOI: 10.1016/S0005-2736(99)00177-7

Pedahzur, R., B. Polyak, et al. (2004). "Water toxicity detection by a panel of stress-responsive luminescent bacteria." *J. Appl. Toxicol.* **24**: 343–348. DOI: 10.1002/jat.1023

Ptitsyn, L. R., G. Horneck, et al. (1997). "A biosensor for environmental genotoxin screening based on an SOS *lux* assay in recombinant *Escherichia coli* cells." *Appl. Environ. Microbiol.* **63**(11): 4377–4384.

Rettberg, P., C. Baumstark-Khan, et al. (1999). "Microscale application of the SOS-*LUX*-TEST as biosensor for genotoxic agents." *Anal. Chim. Acta* **387**: 289–296. DOI: 10.1016/S0003-2670(99)00049-5

Schweigert, N., S. Belkin, et al. (1999). "Combinations of chlorocatechols and heavy metals cause DNA degradation *in vitro* but must not result in increased mutation rates *in vivo*." *Environ. Molec. Mutagen.* **33**: 202–210. DOI: 10.1002/(SICI)1098-2280(1999)33:3%3C202::AID-EM4%3E3.0.CO;2-C

Song, Y., G. Li, et al. (2009). "Optimization of bacterial whole cell bioreporters for toxicity assay of environmental samples." *Environ. Sci. Technol.* **43**(20): 7931–7938. DOI: 10.1021/es901349r

Storz, G. and J. A. Imlay (1999). "Oxidative stress." *Curr. Opin. Microbiol.* **2**(2): 188–194. DOI: 10.1016/S1369-5274(99)80033-2

van der Lelie, D., L. Regniers, et al. (1997). "The VITOTOX test, an SOS bioluminescence *Salmonella typhimurium* test to measure genotoxicity kinetics." *Mutation Res.* **389**(2–3): 279-290. DOI: 10.1016/S1383-5718(96)00158-1

Van Dyk, T. K., W. R. Majarian, et al. (1994). "Rapid and sensitive pollutant detection by induction of heat shock gene-bioluminescence gene fusions." *Appl. Environ. Microbiol.* **60**(5): 1414–1420.

van Dyk, T. K., T. R. Reed, et al. (1995). "Synergistic induction of the heat shock response in *Escherichia coli* by simultaneous treatment with chemical inducers." *J. Bacteriol.* **177**(20): 6001–6004.

Van Dyk, T. K., D. R. Smulski, et al. (1995). "Responses to toxicants of an *Escherichia coli* strain carrying a *uspA'::lux* genetic fusion and an *E. coli* strain carrying a *grpE'::lux* fusion are similar." *Appl. Environ. Microbiol.* **61**(11): 4124–4127.

Vollmer, A. C., S. Belkin, et al. (1997). "Detection of DNA damage by use of *Escherichia coli* carrying *recA'::lux, uvrA'::lux,* or *alkA'::lux* reporter plasmids." *Appl. Environ. Microbiol.* **63**(7): 2566–2571.

Yagur-Kroll, S., B. Bilic, et al. (2009). "Strategies for enhancing bioluminescent bacterial sensor performance by promoter region manipulation." *Microb. Biotechnol.* **doi:10.1111/j.1751–7915.2009.00149.x**. DOI: 10.1111/j.1751-7915.2009.00149.x

Yura, T. and K. Nakahigashi (1999). "Regulation of the heat-shock response." *Curr. Opin. Microbiol.* **2**(2): 153–158. DOI: 10.1016/S1369-5274(99)80027-7

APPENDIX D

Example Bioreporter Protocols

D.1 QUANTITATIVE ARSENITE MEASUREMENTS WITH AN *E. COLI* LUXAB LUCIFERASE BIOREPORTER

Principle of the test

In the presence of arsenate (AsV) and arsenite (AsIII) the arsenic bioreporter cells *Escherichia coli* DH5α-1478 (pJAMA8-arsR) produce bioluminescence that can be easily measured with a luminometer.

Material

Luminometer

Glass vials (4 mL)

Rotary shaker (30°C)

Solutions

Arsenite stock solution (1000 mg As(III)/L ≈ 50 mM)

sterile Luria-Broth (LB)

Ampicillin stock solution (50 mg/mL), sterilised by filtration through a 0.2 μm sterile filter

n-decanal solution (18 mM in 1:1 v/v ethanol-water)

Sterile glycerol (87% (vol/vol))

Cultivation and storage of bioreporter cells

Plate *Escherichia coli* DH5α (pJAMA8-arsR) on LB agar supplemented with 50 μg/mL ampicillin. Incubate overnight at 37°C.

Pick one colony and inoculate 5 mL LB supplemented with 50 μg/mL ampicillin. Incubate overnight at 37°C.

Inoculate 50 mL LB with 1 mL of the overnight culture. Incubate at 37°C with shaking at 200 rpm until an optical density of ≈ 0.6 (wavelenght 600 nm).

Mix the whole culture (50 mL) with 10 mL cold and sterile glycerol (87% (vol/vol)).

Divide into aliquots of 0.65 mL in sterile Eppendorf vials and store them at -80°C.

Bioreporter assay

Prepare a calibration series with 0, 0.1, 0.2, 0.4, 0.8, and 1.0 μM AsIII in water;

Thaw frozen stocks of bioreporter cells for 2 min at 30°C and dilute them in LB (1.3 mL bioreporter cells in 10 mL LB);

Mix 0.5 mL of this bioreporter suspension with 0.5 mL of water sample in a 4 mL glass vial suitable for the luminometer;

Incubate any unknown water samples to be tested for arsenic contamination in the same manner;

Cover vials with screw-cap and incubate them on a rotary shaker at 200 rpm and 30°C for at least 30 minutes;

Add 50 μl n-decanal solution to each vial;

Mix and incubate for another 3 minutes;

Measure light emission in the luminometer. Lighte emission is expressed as relative light units (RLU).

Data handling

The concentrations of inorganic arsenic in the unknown samples are interpolated from a calibration curve.

Remarks

1) The intensity of the bioluminescence is proportional to the arsenic concentration in the range of 7.5 to 75 μg AsIII/L (= 0.1 to 1 μM As).

2) The bioreporter cells also react to arsenate (AsV) at about 10% of the response to arsenite (AsIII).

3) In stead of 4 ml glass vials, 96-well plates can be used for the assay. In that case, the assay volume is scaled down to 200 μl. Relative proportions of cells and chemicals remain the same.

4) Water samples may have to be pretreated when they contain large amounts of iron. See: Trang et al. [2005].

REFERENCES

Stocker, J., D. Balluch, et al. (2003). "Development of a set of simple bacterial biosensors for quantitative and rapid field measurements of arsenite and arsenate in potable water." *Environ. Sci. Technol.* 37: 4743–4750. DOI: 10.1021/es034258b

Trang, P. T., M. Berg, et al. (2005). "Bacterial bioassay for rapid and accurate analysis of arsenic in highly variable groundwater samples." Environ. Sci. Technol. 39(19): 7625–7630. DOI: 10.1021/es050992e 142

D.2   ARSENIC MEASUREMENT USING AN *E. COLI* GFP BIOSENSOR BY EPIFLUORESCENCE MICROSCOPY

Principle of the test

In the presence of arsenate and arsenite the bioreporter cells *E. coli* DH5α-1598 (pProbe-*arsR*-ABS) synthesise GFP (green fluorescent protein). Fluorescence emitted by such cells can be visualised by epifluorescence microscopy. For a quantitative analysis of GFP fluorescence emitted from individual bacteria fluorometry, flow cytometry or digital image analysis have to be used.

Material

Epifluorescence microscope, microscope slides and cover slips

EGFP-filter (emission wavelength ≈ 525 nm)

Greiner 24-well plate

Eppendorf centrifuge

50 mL Greiner tubes

Solutions

Arsenite stock solution (1000 mg As(III)/L, ≈50 mM)

Sterile Luria-Broth (LB)

Kanamycin sulphate (50 mg/mL), sterilised by filtration through a 0.2 μm sterile filter.

Sterile minimal salts medium (MSM; per liter: 0.5 g NaCl, 1 g NH$_4$Cl, 5.5 g MOPS free acid, 5.1 g MOPS sodium salt, 0.05 g Na$_2$HPO$_4$·2H$_2$O, 0.045 g KH$_2$PO$_4$, 0.02% glucose, 20 mmol MgSO$_4$, 1 mmol CaCl$_2$, pH7. MOPS = 3-(N-morpholin-o)propanesulfonic acid)

Preparation of bioreporter cell cultures

Plate *Escherichia coli* DH5α (pProbe-*arsR*-ABS) on LB agar supplemented with 50 μg/mL kanamycin sulphate. Incubate overnight at 37°C;

Pick one colony and inoculate 5 mL LB supplemented with 50 μg/mL kanamycin sulphate. Incubate for 14 hours (overnight) at 37°C on a shaking platform;

Dilute 0.4 mL overnight culture in 20 mL LB supplemented with 50 μg/mL kanamycin sulphate;

Continue incubation for about 3 h at 37°C until turbidity of the cell suspension is ≈ 0.6 (wavelenght 600 nm);

Centrifuge 20 mL of the suspension in a 50 mL Greiner tube for 5 min. and at room temperature at 4'500 rpm;

Decant supernatant;

Resuspend cell pellet in 20 mL MSM by slowly vortexing.

Bioreporter assay (epifluorescence microscopy)

Prepare a calibration series with 0, 0.5, 1.0, and 5.0 μM AsIII in water;

Assay mixtures contain: 0.5 mL diluted cell suspension (in MSM) and 0.5 mL sample (or arsenite calibration solution). Mix in Eppendorfs or wells of a 24-well plate;

Incubate at 37°C with shaking, if possible;

Sample at 1h, 2h and 4h.

GFP quantification per epifluorescence microscope

Transfer 1 μl of the assay mixture to be analysed to a microscope slide. Place a cover slip and observe at 400-1000 fold magnification with a proper filter for EGFP;

If necessary to concentrate the amount of cells visible in the microscope image, centrifuge a sample of 0.2 mL for 1 min at 13,000 rpm;

Decant supernatant and resuspend in 10-50 μl of MSM;

Transfer 1 μl to a microscope slide and place coverslip.

Data handling

Fluorescence intensities can be compared to a negative control (no arsenic added) and the positive controls by 'visual inspection';

For quantitative analysis it is necessary to make standardized images, e.g., with a digital camera and fixed exposure times (100 - 500 msec);

The brightness of the objects in the digital image file can subsequently be analysed with specialised software, such as ImageJ or NIHImage (freeware), or Metaview (Image Corporation).

Alternative reporter assay using fluorometry

Prepare a calibration series with 0, 0.1, 0.2, 0.4, 0.8, 1.0, and 2.0 μM AsIII in water;

Assay mixtures contain: 100 μl diluted cell suspension (in MSM) and 100 μl sample (or arsenite calibration solution). Mix in wells of a 96-well plate;

Incubate at 37°C with shaking, if possible;

Measure Absorbance (e.g., at 600 nm) and Fluorescence (at 525 nm) in a Fluorimeter after 1, 2 and 4 h;

Correct Fluorescence values for culture turbidity;

Plot corrected fluorescence values versus arsenite concentrations and calculate arsenite equivalent concentrations in the unknown samples by interpolation.

Remarks

If MOPS is unavailable, one can replace it by a phosphate buffer (20 mM, pH 7).

In this protocol, glucose is essential for providing energy to the cells.

See above for luciferase protocol: pretreatment is necessary for water sample with high iron content.

REFERENCES

Wells, M., M. Gösch, et al. (2005). "Ultrasensitive reporter protein detection in genetically engineered bacteria." *Anal. Chem.* 77: 2683–2689. DOI: 10.1021/ac048127k

D.3   ARSENIC MEASUREMENTS WITH AN *E. COLI* BETA-GALACTOSIDE BIOREPORTER

Principle of the test

In the presence of arsenate and arsenite the arsenic bioreporter cells *E. coli* DH5α -1595 (pMV-*arsR*-ABS) synthesise the enzyme beta-galactosidase. The activity of beta-galactosidase can be quantified per spectrophotometer assay (Miller assay with ONPG), per potentiometric assay using PNPG as substrate, or observed visually by using X-Gal as substrate. X-Gal is converted by the enzyme to a blue colour. The intensity of the blue colour is a measure for the exposure of the cells to arsenic.

This is the *simplest* version of the arsenic biosensor test.

Materials

Sterile tooth picks

24-well Greiner plates (or equivalent transparent incubation vials)

Solutions

Arsenite stock solution (1000 mg As(III)/L $\approx$ 50 mM)

Sterile Luria-Broth (LB)

Ampicillin stock solution (50 mg/mL), sterilised by filtration through a 0.2 μm sterile filter

X-Gal solution (25 mg/mL in dimethylformamide, store at -20°C in the dark)

Preparation of bioreporter cell cultures

Plate *Escherichia coli* DH5α (pMV-arsR-ABS) on LB agar supplemented with 50 μg/mL ampicillin. Incubate overnight at 37°C;

Pick one colony and inoculate 5 mL LB supplemented with 50 μg/mL ampicillin. Incubate for 14 hours (overnight) at 37°C on a shaking platform;

Dilute 0.4 mL overnight culture in 20 mL LB;

Transfer 2 mL of this diluted cell suspension to a new vial and add 0.1 mL X-Gal solution. This is the cell suspension to be used in the assay.

Bioreporter assay

Prepare a calibration series with 0, 0.1, 0.5, 1.0, 2.0 and 5.0 μM AsIII in water;

Assay mixtures contain: 0.1 mL diluted cell suspension (in LB) and 0.9 mL sample (or arsenite calibration solution). Mix in Eppendorfs or wells of a 24-well plate;

Incubate at 30-37°C;

Inspect the colour development after 3-5 hours of incubation. (The assay can be incubated for as long as the negative control remains colourless.)

Interprete the intensity of the colours in unknown samples compared to the blanc (no arsenic) and to the 0.1 (= 7.5 μg As/L) and 0.5 μM (= 37.5 μg As/L) calibration values (corresponding to the most common arsenic drinking water standards).

REFERENCES

Wackwitz, A., H. Harms, et al. (2008). "Internal arsenite bioassay calibration using multiple reporter cell lines." *Microb. Biotechnol.* 1(2): 149–157. DOI: 10.1111/j.1751-7915.2007.00011.x

D.4   SAMPLE PRETREATMENT

D.4.1   WATER

Background

Water samples with an iron content above 0.5 mM can complex arsenic in solution, which renders it unavailable for the bioreporter cells. The pretreatment uses acidification down to pH 2, after which the iron is complexed with pyrophosphate.

Solutions

200 mM sodium pyrophosphate solution ($Na_4P_2O_7,10H_2O$, Sigma))

HNO_3

4 ml glass vials

Treatment and assay

Prepare bioreporter cell suspension from frozen stocks (1.3 ml frozen aliquot with 10 ml sterilized LB medium);

Acidify the water sample with HNO_3 to pH 2 (final concentration of HNO_3 = 0.015 mM);

Mix acidified sample with bioreporter suspension in LB in a 1:1 (*v/v*) ratio;

Resulting pH should be at least 5.5 (to be checked);

Immediately add 25 μl of 200 mM $Na_4P_2O_7$ solution (final concentration 5 mM). pH should be 7.

Incubate assay at 30°C for 90 min and at 200 rpm

Add 50 μl of 18 mM *n*-decanal to measure luciferase activity

D.4.2   RICE

Background

Not only aqueous samples, but also food stuffs can contain (large) amounts of arsenic. Part of this arsenic is in form of organic arsenicals; a significant part is biologically available as arsenite or arsenate. To release arsenite and arsenate from the food and bring it into aqueous solution, an enzymatic pretreatment is necessary.

Materials

Glass beads (ø 0.1 mm)

Fastprep cell disruptor

Water baths

Solutions

15 mM HNO_3

Sørensen's phosphate buffer solution (0.67 M, pH 7.4)

Fresh pancreatin solution (200 mg pancreatin per 1.5 ml)

Extraction

Mix 0.1 g rice flour, 0.1 g glass beads, and 1.4 ml HNO_3 [15 mM] in 2 ml-screw cap tubes;

Vortex twice in Fastprep (speed 6.5, time 45 s); cool on ice, vortex twice in Fastprep (speed 6.5, time 45 s);

add 0.1 ml pancreatin solution

incubate for 8 h at 37° C with shaking;

Vortex and incubate for 16 h at 55°C, for example in a hybridization oven;

Centrifuge for 1 min at 13,000 rpm and room temperature

transfer 600 μl supernatant to a clean tube, discard remaining supernatant;

add again 1 ml HNO_3 [15mM] to the pellet

extract again for 3 h at 55°C

centrifuge for 1 min at 13,000 rpm

transfer 0.6 ml supernatant to a second clean tube

Inactivation of enzyme

Place tubes with supernatant in a boiling water bath for 15 minutes;

Neutralization

add 10 vol% Sørensen's phosphate buffer solution to the supernatant of each tube (e.g., 70 μl to 0.7 ml supernatant);

Bioreporter assay (for example with the luciferase assay)

thaw four aliquots of frozen strains (4 x 0.65 ml cell suspension)

centrifuge 1 min for 13,000 rpm

remove supernatant

resuspend each cell pellet in 0.4 ml LB and pool all four suspensions (final volume: 1.6 ml)

Mix in a well of a 96-well plate: 170 μl rice treated sample, 10 μl bioreporter cell suspension, 20 μl 10-times concentrated LB medium;

Prepare calibration series as previously but using HNO_3 and Sørensen buffer;

Incubate for 30 min to 1 h; add n-decanal and measure luciferase activity.

REFERENCES

Baumann, B. and J. R. van der Meer (2007). "Analysis of bioavailable arsenic in rice with whole cell living bioreporter bacteria." *J. Agric. Food Chem.* 55(6): 2115–2120. DOI: 10.1021/jf0631676

Trang, P. T., M. Berg, et al. (2005). "Bacterial bioassay for rapid and accurate analysis of arsenic in highly variable groundwater samples." *Environ. Sci. Technol.* 39(19): 7625–7630. DOI: 10.1021/es050992e

Author's Biography

JAN ROELOF VAN DER MEER

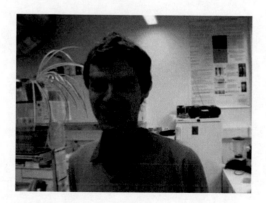

Jan Roelof van der Meer is Associate Professor in Environmental Microbiology at the Department of Fundamental Microbiology of the University of Lausanne, Switzerland. He joined the University of Lausanne in 2003, after spending ten years as Group Leader at the Swiss Institute for Aquatic Science and Technology (Eawag). He completed a MSc degree in Environmental Sciences from the Wageningen Agricultural University (The Netherlands), and holds a PhD degree of the same university specializing in molecular microbiology. Before joining Eawag he was postdoctoral fellow at the Dutch National Dairy Institute. His primary field of interest concerns the manyfold interactions of bacteria with chemical pollutants in the environment. In ongoing research his group actively pursues the evolutionary mechanisms underlying adaptation of bacteria to using organic pollutants as unique carbon and energy sources. Another part of his research focuses on pollutant degradation by bacteria in the environment. A third activity of his group concentrates on the design, construction and application of bacterial bioreporters for environmental quality measurements, which is the topic of this lecture. Dr. van der Meer coordinates the FP7 large integrated European project BACSIN on bacterial survival and adaptation in the environment. Before that he served as coordinator of the FP7 project FACEiT, which focused on biology-based detection tools for environmental quality assessment.

Printed in the United States
by Baker & Taylor Publisher Services